Doormaking

Materials, Techniques, and Projects
for Building Your First Door

Strother Purdy

LINDEN PUBLISHING
FRESNO

Doormaking
Materials, Techniques, and Projects
for Building Your First Door
by Strother Purdy

Cover design: Troy Thorne
Artwork: Jim Goold
Art direction and layout: Maura J. Zimmer
Photography: Strother Purdy except where noted.

ISBN: 978-1-610352-91-8

579864

Linden Publishing titles may be purchased in quantity at special discounts for educational, business, or promotional use. To inquire about discount pricing, please refer to the contact information below. For permission to use any portion of this book for academic purposes, please contact the Copyright Clearance Center at www.copyright.com

Printed in China.

Woodworking is inherently dangerous. Your safety is your responsibility. Neither Linden Publishing nor the author assume any responsibility for any injuries or accidents. Photographs in this book may depict the usage of woodworking machinery where the safety guards have been removed. The guards were removed for clarity. We urge you to utilize all available safety equipment and follow all recommended safety procedures when woodworking.

Library of Congress Cataloging-in-Publication Data

Names: Purdy, Strother, author.
Title: Doormaking : materials, techniques, and projects for building your
    first door / Strother Purdy.
Description: Fresno : Linden Publishing, [2017] | Includes index.
Identifiers: LCCN 2017001604 | ISBN 9781610352918 (pbk. : alk. paper)
Subjects: LCSH: Wooden doors. | Doors.
Classification: LCC TH2278 .P887 2017 | DDC 694/.6--dc23
LC record available at https://lccn.loc.gov/2017001604

The Woodworker's Library®

Linden Publishing, Inc.
2006 S. Mary
Fresno, CA 93721
www.lindenpub.com

Door, *noun*:
1. A movable barrier of wood or other material, usually turning on hinges or sliding in a groove, and serving to close or open a passage into a building or room.
2. That which the dog is always on the wrong side
—*Oxford English Dictionary*, mostly.

Dedicated to my children, Josephine and Isaac, for no particular reason other than that I love them very much, perhaps with the hope that they will eventually stop slamming the front door. That door that I made, you know. That's really nice. And shouldn't be slammed so hard. Not while I'm in earshot anyway.

Acknowledgments

David Brothers and Rebecca Cheng
Jeff Cook
Barbara Dahl
Bill Duckworth
Andy Engel
Glen Hochstetter
Jon Lindblom
Joseph Manley
Jon Olivieri
Marc Olivieri
Geoffrey Purdy
David and Nancy Sposato
Eric Vikstrom
Laurie Wesley

The Connecticut Office of the State Building Inspector
Conway Hardwoods
Atlantic Plywood
Historic Housefitters
H. H. Taylor's Hardware
Ring's End
Whitechapel, Ltd.

Safety Tip

Have you read those endless pages of safety warnings that come with every new power tool? You know, the ones up front in the user manual that go on and on and on about everything obvious (do not swing running power tool by its cord above your head, etc.) that you're supposed to read thoroughly before taking the tool out of the box? They're written by the underpaid employees of rich lawyers as part of a lucrative settlement when some poor person did swing the tool around and lost an ear thereby winning millions of dollars in court to be given to the lawyer.

Don't be this person, which is to say that keeping your digits attached is much better than spending long hours without them in a courtroom. What's my advice on how to keep all your digits? The lawyer will tell you it's all those safety warnings. I'll add that it's important to listen to your intuition. This is that gut feeling that says "this isn't right" or "I'm not sure about this" or "I hope this works ok." If the tablesaw scares you, it's for a very good reason. Leave it alone and use other tools. If the tool doesn't act the way it should, there's a reason. Stop and ask advice from someone who knows. Learn about the tool or technique and only use it on your own at the point that it no longer scares you.

Contents

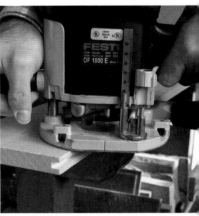

Then and Now

I have a copy of *Henley's Twentieth Century Book of Recipes, Formulas and Processes: Containing Nearly Ten Thousand Selected Scientific, Chemical, Technical and Household Recipes, Formulas and Processes for Use in the Laboratory, the Office, the Workshop and in the Home*, published in New York in 1907. Intended for the general public, Henley's tells you how to make everything, from your own acid-proof wood finishes to food preservatives, foot powders, even zinc contact silver plating. But in close to eight hundred densely packed pages, there is not a single image or illustration, and there are few how-to instructions. To make the acid-proof wood finish, for example, the book simply tells you to boil copper sulfate and potassium chlorate salts until dissolved.

To many modern Americans, the directions in this book are dangerously incomplete, perhaps criminally. Will boiling these crystals give off noxious fumes? Do you add water? How much? Can they be boiled safely in a steel or aluminum pot? Where can you even buy these crystals? Henley's is mute. While I'm sure many readers in 1907 did not know the answers to all these questions, many more did. Henley's was a popular book.

In 2015, we no longer have the same foundation of practical knowledge or hand skills (or understanding of risk) that our ancestors did. The efficiencies of mass manufacturing has allowed each of us to focus on specialized professions—we simply don't need to build or repair our own houses or furniture anymore; there's someone else who specializes in that. Many of us lament this loss, looking back on earlier generations, amazed by their ability to do so much on their own.

But they did not have the internet, heated toilet seats or inexpensive Asian manufacturing integrated with global markets and high volume shipping capacities. With these things, we are far more productive, far less burdened by the work necessary to simply survive, and have far greater freedom to pursue other opportunities. We are now far less independent and far more interdependent. We are not worse off—we are just different.

Today, most people would think it slightly insane to make your own doors when you can buy a manufactured, pre-hung steel and plastic door at Lowe's for $67. That's not much less than three times the cost of this book.

Of course you won't save any money making your own doors, or advance your career in selling widgets. You will, however, create beautiful, unique and useful objects—thereby acquiring the traditional skills of craftsmanship that have shaped our humanity: a main advantage, and consequence of the opposable thumb.

If we share anything with our handier ancestors from 1907, it's the need for understanding and mastery of our daily environment. We may live in a world where merely opening the back of a phone voids the warranty, but we still can find deep satisfaction in creating some of the things we use.

Perhaps you have long experience working with your hands. Then enough said—this book will give you answers to your questions about making a door that won't warp, crack or fail unexpectedly. But perhaps you're new to the idea of making your own things, and a book seems like the right place to start. There are simple doors you can make and you'll find them here, with how-to instructions shown step-by-step to help you make them right the first time.

And if you get a copy of Henley's, you will know how to finish a door so it's acid-proof—perfect for the Zombie Apocalypse, because, as we all know, by next year the TV zombies will know how to spit acid, if they don't know already.

The Basic Challenges of Doormaking: the Origin of Traditions

In caveman times, men sat around playing with the fire while women did all the work. This was noticed, and so the first honey-do project was devised. Some might suggest that project was a table or a stool, or even a bowl. But I think the front door was man's first honey-do project.

"The cave is drafty, dear. And bears wander in too often. Why don't you put down that flaming stick and make us a movable barrier that will serve to either open or close the way into our cave."

"Durrrrrr," our common ancestor stalled for time to think his way out of this task.

"Whatever you want to call it, dear. Just get busy."

"Yes dear."

At this moment, the concept of the door was born—at one stroke solving the problems of keeping cold/animals/insects/nosy neighbors out of our homes, and heat/possessions/wandering toddlers/family secrets in our homes.

Now our common ancestor racked his brain to bring this great concept to life. What to use? Straw and animal skins might work, but were too flimsy—bears could easily paw or gnaw through them. Stone would stop a bear cold, but was too heavy and hard to work. That left wood: a durable material that wasn't too heavy and was relatively easy to work, and yet would keep out the bears.

Carving a door out of a single slab of green wood to fit the cave entrance seemed like the easiest approach. And so our common ancestor tried it. But that door warped like a potato chip (though those hadn't been invented yet—that would occur much later, in Greece), refusing to shut after the first week.

Nice property, but how to keep out the bears?

For his second door, he seasoned the plank beforehand, letting it dry for several years. It was more stable, but it still grew and shrunk depending on the season, tended to cup, stuck shut in the summer and left a gap in the winter. Then splits developed at the ends and it eventually broke itself apart.

For his third through 50th door, our common ancestor tried a variety of designs to solve an ever growing list of design challenges—and through this process understood the basics of making doors with wood. On the one hand, wood has a lot going for it:

- **Wood is relatively easy to shape (with tools)**
- **Wood is relatively light compared to its strengths**
- **Wood is a durable material**
- **And it looks pretty good**

And on the other, wood has a few core problems that have to be accommodated:

- **Wood is weak across the grain**
- **Wood grows and shrinks according to the relative humidity**
- **Certain woods get eaten by fungus and insects**
- **Without a finish, wood silvers and degrades with UV light, oxygen and water**

Discovering that wood is a reasonably complex material with a range of attributes that vary from species to species, tree to tree and even inch by inch kept our common ancestor pretty busy. Then consider that an exterior door will face winter on one side, and a warm interior on the other—huge temperature and humidity differences that stress the wood like nothing else. It quickly became evident that doors had a lot of difficult, even unique challenges. Hard-won solutions came with time and effort.

So before our first doormaking common ancestor died, he passed along a lot of valuable advice on making doors to his apprentice, who understood half of it because he was a teenager and not listening. But when that teenager grew up, he still made doors the same way because making them differently took more time and effort and they broke faster. This was the beginning of **Tradition**—the art of doing it the same way it has always been done because it works.

Practical Design Principles for Doors

While we can only imagine when and where the first solid wood doors were made, we certainly have been making them since long before the Egyptians walked around in profile. The amazing thing is just how similar wood doors are all over the world (including cabinet doors, gates and other variations): people everywhere largely reached the same engineering solutions and refined them on the anvil of everyday life to create traditions that have persisted, despite superficial stylistic differences. In short, there are good reasons why doors look like they do and have looked much the same for a long time.

Of course there's a place for innovation and creativity—all doors do not look the same as tradition has width and depth. Within traditions are stylistic innovation and experimentation. However, we forget why we do things, so we make new mistakes to learn old lessons. The more you understand the reasons behind traditions, the more likely you'll be successful when you make something new and unique.

If you're just here to build one of the projects like a recipe, I'll caution against it and recommend you read and absorb the whole book first. If you want to make doors on your own without following any of the projects, this chapter will help you understand the practical aspects of doormaking traditions, materials, hardware and building technique choices.

At the Fork in the Road, Go Both Ways—Answering the Age Old Question "What's the *Best* Way to Do It?"

When I was young and starting out in the profession, everything was simple and easy. I knew of one good way to cut joinery, three good woods to use and one good finish. I had the answers.

Problems and failures, eventually taught me otherwise. Over time I learned more techniques, woods, finishes and other details that helped to get things right more often than before. Often I felt I found the "best way" after trying ten ways. But then the next problem argued against it, and I'd look for an eleventh.

Experience leads to a more refined decision making process: answers generally begin with a "that depends" and then a mulled solution that will "probably work best." It doesn't produce "The Best Way." That doesn't exist, and I don't have it, no matter what I claim, as there is always tomorrow to prove the sum of history wrong.

For the beginner looking for a clear direction, projects with step-by-step instructions are just the ticket: they are a distillation of experience, choices made for you that should get you to a successful door if followed faithfully. Experience isn't really transferable, but it can be created through doing. When you start making doors to your own design, without the help of step-by-step instructions, the basic principles in this section can serve as guides to minimize those pesky problems and failures.

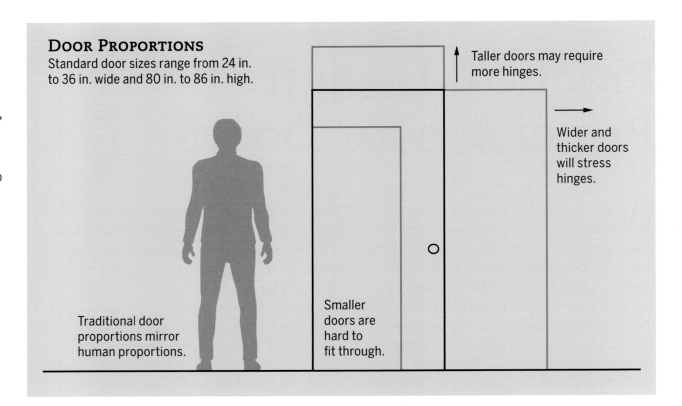

DOOR PROPORTIONS
Standard door sizes range from 24 in. to 36 in. wide and 80 in. to 86 in. high.

Taller doors may require more hinges.

Wider and thicker doors will stress hinges.

Traditional door proportions mirror human proportions.

Smaller doors are hard to fit through.

Build doors to be rectangular, taller than they are wide, and relatively thin.

The tall rectangle is the basic shape of the human body and has been the basic shape of the door for ever (though with growing global obesity, we might be moving towards a square or circular door tradition).

Modern manufactured doors come in a few standard sizes—for example, 30 in. by 80 in. and 36 in. by 80 in. These are good proportions to stick with as they work, are within code, and look right. However, the beauty of a custom door is that you can make it however you like.

The maximum width I'd make a single door is a about 40 in., as adding width adds weight and stress on the higher hinges, causing them to bind and fail faster than they should. For doorways wider than that, I'd make a pair of doors that close in the middle.

Doors don't need to be taller than 80 in. for any practical reason, unless you have a basketball player in the family or plan on building sailboats in your dining room that need a way out. That said, don't think of 80 in. as a limit—wider doors tend to look better if they are proportionally taller. A grand double entry door I recently completed was 56 in. wide by 86 in. high. None of these good proportions are based on the golden rectangle. Sorry to disappoint the classicists.

Moving outside of the tall rectangle rule will add complication and impracticality, but perhaps add some interest. Curved top doors are practical and can be beautiful, but are much more difficult to make—remember that you have a matching curved jamb. Circular doors are cute and may remind you of Hobbits, but are a nightmare to hang as there's one point for them to hinge against, and just think of the lawsuits you'll have to fend off from guests tripping on the curved jamb near the floor.

The third dimension of a door—its thickness—is no less important. Wood doors should range between 1¼ in. to 2 ¾ in. thick. I make 36 in. by 80 in. doors on average about 2 in. thick. If you make

Consider Who Will Use the Door

I recall wondering why a set of doors on a 17th century castle were so tall and wide—about 12 ft. high but only 5 ft. wide. People were supposed to be smaller back then, right? Later on the tour, the guide explained that the Lord liked to ride his horse inside, and so commissioned those special doors. The practical lesson is that if you're building a door to let something other than people through it, consider their proportions and design accordingly.

More down to earth, for example, do you have a friend in a wheelchair? Make sure your doors are wide enough to let a standard-width wheelchair get through, including arms and elbows, without a problem, and offer enough maneuvering space around it to both open and close it. The Americans with Disabilities Act (ADA) does not apply to residential homes, but that doesn't mean you can't incorporate some of their guidelines into your doors. Also consider the location of doors relative to each other—when they open do they get in the way of each other? You can find detailed ADA guidance on handle and knob heights, opening clearances and other issues in the most recent ADA Standards for Accessible Design, available as a PDF here: http://www.ada.gov/regs2010/2010ADAStandards/2010ADAStandards_prt.pdf

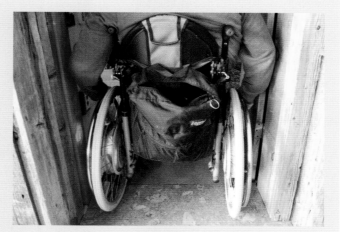

Wheelchairs need extra width between jambs to pass without damage.

them much thicker, they'll get too heavy and hard on their hinges. Make them much thinner, and they will be more likely to warp and break. The thinnest doors tend to be screen doors, largely because they have to share a jamb with a full-thickness door.

Locating Hinges and Handles

Hinges and handles (or knobs) can go many places on a door, but not anywhere. I put a typical handle approximately 36 in. off the floor, and any locks right above it. It's an easy number to remember—3 ft.—and is comfortable to reach for me (I'm 6' 0" tall) and my wife (she's 5' 0"). No grown adult has to stoop or reach for this handle, and yet toddlers can't reach it to give themselves unapproved freedom. This location also falls within the range recommended by the American Disabilities Act (ADA) Door Hardware Mounting Height Range, which is between 34 in. and 48 in.[1]

The number and location of hinges is a little more complex and depends on several factors, including door height, weight, width and type of hinge. Happily, most door hinges come with spec sheets recommending number and spacing for the size and weight of door you're hanging. For example, 5 in. ball-bearing hinges will carry more weight than simpler 3½ in. hinges, so you would need fewer of them for a given door.

---

1. (International Building Code 2009: 1008.1.9.2)

The rule of thumb for a 24 in. to 36 in. wide and 80 in. tall door is three hinges, spaced 5 in. to 7 in. from the top, 9 in. to 11 in. from the bottom and the third spaced evenly between the other two. For heavier or taller doors, you might add another hinge, or use ball-bearing hinges that can carry more weight. For smaller, lighter doors, you might be able to get away with only two hinges.

Accommodate Wood Movement in the Design

As our common ancestor found out on Day 2—wood moves. Any durable and effective door design has to take wood movement into consideration of construction or it will break itself apart. The three basic forms construction approaches are the board-and-batten, frame-and-panel and solid/hollow core doors. Each addresses wood movement in a different way.

The simplest solution is the board-and-batten, though it is not sturdy and can't be easily weatherproofed. It stays together, but is not dimensionally stable. The pinnacle of solid-wood technology is the frame-and-panel door. It is both durable and dimensionally stable. The third is the most modern: hollow and solid core doors that use manufactured sheet goods and adhesives rather than solid wood. These doors are strong, dimensionally stable, but don't look traditional.

HANDLE AND HINGE LOCATIONS

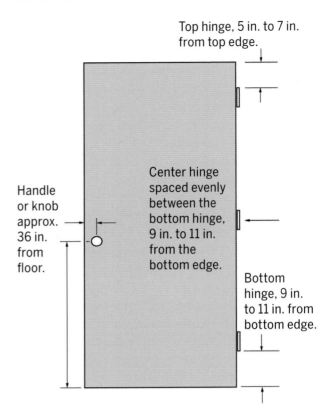

Top hinge, 5 in. to 7 in. from top edge.

Handle or knob approx. 36 in. from floor.

Center hinge spaced evenly between the bottom hinge, 9 in. to 11 in. from the bottom edge.

Bottom hinge, 9 in. to 11 in. from bottom edge.

Build Doors to Last

The three primary sources of wear and tear on doors are people, people and people. Sure, beasties can eat the wood in your doors, and the weather can be pretty brutal, but your main adversary is people—teenagers in particular. No one can slam a door with such

Why Closer to the Top?

"Just do it that way and stop asking questions!"

When I started out as an apprentice, I wondered why the hinge at the top was closer to the door edge than the bottom. Tradition can sometimes be opaque, as can employers.

But this time it turns out to be simple physics (though I'm sure I'll get a thirty page rebuttal from an actual physicist on how I'm only partly right on this, and using all the wrong terminology)—the more door weight above a hinge, the more racking

force the door puts on it, the shorter life the hinge will have, and the better chance you have of the door sagging. The hinge at the bottom is therefore doing less work (really a different kind) than the one at the top. Placing a hinge at the very top of the door edge might resist racking best, but that location may conflict with the joinery and doesn't look good. So the tradition is a slight compromise: to raise all three hinges just a bit from perfectly even spacing on the edge.

BOARD-AND-BATTEN

Vertical boards held together by horizontal battens give you the simplest of solid wood doors and are somewhat dimensionally stable.

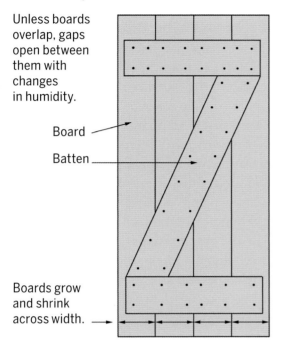

Unless boards overlap, gaps open between them with changes in humidity.

Board

Batten

Boards grow and shrink across width.

FRAME-AND-PANEL

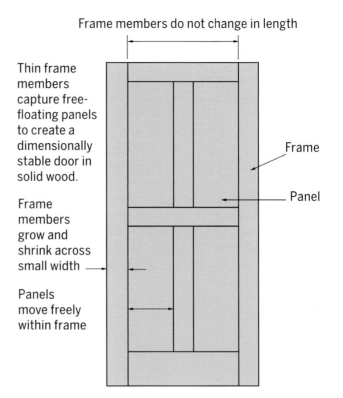

Frame members do not change in length

Thin frame members capture free-floating panels to create a dimensionally stable door in solid wood.

Frame members grow and shrink across small width

Panels move freely within frame

Frame

Panel

destructive violence as an angry teenager. Build your doors to survive these challenges.

I've never worried much about bugs or fungus eating my doors unless they're exterior. Many wood species, especially maple, contain a lot of food energy (starches, sugars, etc.) for enterprising animals and plants capable of eating it. This includes a wide range of insects such as boring beetles and microscopic life such as fungi. At first, it will produce added interest to the grain (spalting), but unchecked, bugs and fungus will eventually reduce your door to dust. It's not a common problem, however, as interior doors rarely get wet enough to foster fungal growth or be habitable for the vast majority of insects (just pick apart a rotting log in a forest sometimes to see the amazing variety of what lives and thrives in wet wood). For an exterior door that does get wet and could stay wet, it's best to use a wood that is naturally resistant to rot (discussed in the next chapter).

HOLLOW AND SOLID CORE

Modern sheet materials skinned over a wood frame with modern adhesives create a durable and dimensionally stable door.

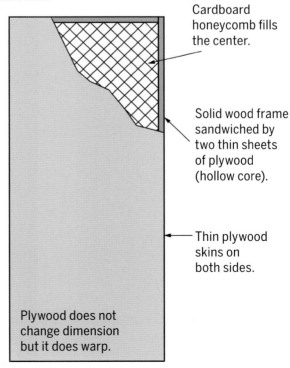

Cardboard honeycomb fills the center.

Solid wood frame sandwiched by two thin sheets of plywood (hollow core).

Thin plywood skins on both sides.

Plywood does not change dimension but it does warp.

Tradition and Creativity

Traditions help us be successful without fully understanding why. They come from many cultural sources; but for the modern woodworker, the most important traditions are the ones based on the characteristics of the materials. Laid down in the aftermath of disaster, these traditions are strategies that keep those disasters at bay. If the master ever understood why those traditions were developed, he sure as hell didn't tell his ungrateful, snot-nosed apprentice. In turn, to become a Master, those Apprentices had to do as the Master said without asking too many annoying questions to avoid a sound beating. Apprentices thus made doors the way their masters showed them, to both avoid pain and because they knew a door made like the master would work and would last.

Creativity is the evil twin of tradition. Nothing can change or get better without it, but the traditions driven by the characteristics of the material are not open to change without serious consequence. Disobey the aspects of tradition based on the inherent rules of the materials, and they will sit up, take notice and frog-march you towards disaster. The key is to innovate with an understanding of the limits of the material.

Old doors look a lot like new doors for good reasons—their design works and lasts.

Weathering is a result of ultraviolet (UV) light from the sun and oxidation that breaks down wood cells, turning the surface a silver-gray color. Finishes protect wood pretty well from this process (their details are discussed in Chapter 2), but have to be renewed as they wear away. If you live in the Northern hemisphere, a Southern-facing exterior door will take the worst beating from the sun. In fact, I wouldn't recommend installing a Southern-facing exterior wood door unless you plan to refinish it more than once a year. Eastern and Western-facing doors don't fare much better having direct sun for half the day. Northern-facing doors will need less maintenance, as will doors that are somewhat sheltered from the elements by an overhang or a screen door.

Weathering and insect damage, however, are not universally a bad thing. They can look quite nice and are even sought after (look for "wormy chestnut" or "Ambrosia maple"—these varieties have simply suffered insect damage—and note the price tag). You may even want to create the weathered look, easily created through sandblasting and vinegar washes (that's another book, maybe covered in Henley's). Water, however, will eventually lead to rot in every wood, so keep your doors dry.

What is a universally bad thing are doors that get broken through use. Human interaction with doors puts a wide variety of stresses on their joinery and their hinges. Doors demand the strongest joinery to stay together as they are constantly receiving small and large shocks: opened, closed, opened, closed, slammed shut, climbed on, hung from, etc.

The construction methods that compensate for wood movement described above are also pretty good at resisting the stress of daily use. Use them. The particular wood you choose for the door and the

STRESSES ON DOORS

Suspended above the floor by a few hinges, door joinery and hardware are constantly under a variety of stresses.

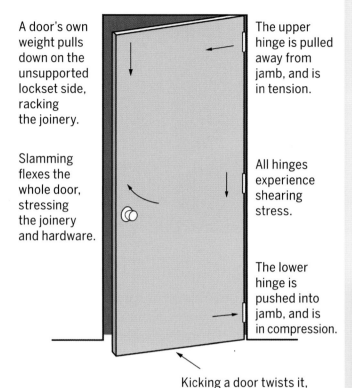

A door's own weight pulls down on the unsupported lockset side, racking the joinery.

Slamming flexes the whole door, stressing the joinery and hardware.

The upper hinge is pulled away from jamb, and is in tension.

All hinges experience shearing stress.

The lower hinge is pushed into jamb, and is in compression.

Kicking a door twists it, racking the joinery.

jamb will also determine how long a door lasts. Species and general wood characteristics are discussed in Chapter 2. And good hardware, especially the hinges, are key to doors that work and don't need maintenance.

Don't Build Doors That Will Kill You—Common Sense, Safety and Building Codes

Did you know that the Consumer Product Safety Commission estimates that 290,287 people were injured by a door in 2014?[2] I didn't. Perhaps to save lives we should all opt for bead curtains between rooms. Then again, maybe not. Still, I think more injuries are done to doors by people, but nobody seems to keep track of that.

Treating Wood for Insects

If you're working with kiln dried wood, you should have no worries about insects. Most wood boring insects can be killed with heat (130 degrees for 4-5 hours). But if you're working with air-dried lumber, you might want to take some precautions. There are many good commercial products for this, but I use a simple solution of Borax. It won't discolor the wood, is about as poisonous to people as table salt, and keeps the beasties at bay.

My method is non-scientific, but seems to work fine. I dissolve about a cup of Borax in a gallon of warm water then brush it on the green boards (Borax will wash off, so don't leave the boards in the rain—anyway, you want them dry, right?).

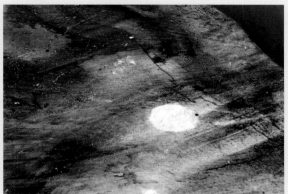

When you see little holes and piles of frass, you have bugs in your wood.

In any case, to help keep you safe (and keep your home energy-efficient), exterior doors are governed by local, state and federal law—yep, there's a building code for them. Interior doors are generally not subject to code, but should be made with the code in mind.

In the United States, the specifications of exterior residential doors are almost universally governed by the 2009 and 2012 International Residential Code (IRC). However, your state, county and town may also have added

---

2. http://www.cpsc.gov/cgibin/NEISSQuery/PerformEstimates.aspx

A standard 36 in. wide door will generally give you a 33½ in. wide opening.

additional restrictions, requirements and variances. Reading the relevant sections in the IRC before you build a door is a good idea (at the time of printing, they were located for free online here: http://publicecodes.cyberregs.com/icod/irc/index.htm).

But read them only if you have a mind for really dense, confusing, technical and bureaucratic language telling you that your doors should, really, be built by an approved manufacturer. (If you dive in, just know that the IRC writers lump doors and windows together as types of "fenestration"—as in "defenestration," which becomes a desirable option for your computer after reading the building code online for an hour or so).

Where does this leave us DIYers? Well, simple. Just speak with your local building inspector before you design and build any exterior doors and he or she will help you through the provisions for DIY doors that apply in your area. It's a conversation well worth having early in the process to avoid the safety and energy conservation problems the code is designed to help you avoid—and a failed inspection.

To help you plan the design of an exterior door, here are a few tips about code issues (This is not an exhaustive summary of door code, or any substitute for the building codes applicable where you live. It's your responsibility to understand the specific codes that apply in your area and adhere to them).

Every residential dwelling must have at least one exterior hinged door with a clear 32 in. wide by 78 in. high opening. Be careful— it's the opening, not the door that needs to be at least 32 in. by 78 in. Consider how far the door will swing out of the way to get true "opening" dimensions. You should be fine if you make your front door the standard 36 in. by 80 in., with standard trim and sill plate, hinges that open beyond 110 degrees and room for the door to do that. Why? You really don't want a smaller opening as it will make it exquisitely difficult to get your new Barcalounger inside and the old one outside, among other large things.

Exterior doors can have a lock that requires a key on the outside, but not on the inside. Why? If you need to get out of your house in a hurry, say to avoid a fire, you can't be looking where you left your keys.

On the other hand, your exterior doors can't be larger than 24 square feet of opening (say 45 in. by 72 in.) without having to be subjected to some complex e-value tests. Why? Too large an opening can make for poor energy efficiency, and a really heavy door too liable to break its hinges, for that matter.

Use tempered, insulated glass panels in both your interior and exterior doors. Why? When tempered glass breaks, it shatters into thousands of small, mostly harmless fragments. Glass that isn't tempered can break into large, sharp shards. So when your door gets slammed too hard by an angry teenager, the harm is only emotional. And the insulation is for energy efficiency and noise reduction.

For attached garages, the door from your house to your garage is an exterior door, and therefore governed by code. It also has some special requirements to make them relatively fireproof. If they're made of solid wood, they have to be at least 1⅜ in. thick at all points. They also need to be self-closing and self-latching. Why? Consider where you keep vehicles and what they have in their gas tanks.

Finally, exterior doors are required to have weather-stripping. Why? Energy efficiency—the government doesn't want you burning money unnecessarily on air conditioning and heating.

SOME COMMON CODE GUIDELINES FOR EXTERIOR DOORS

Weatherstripping is required.

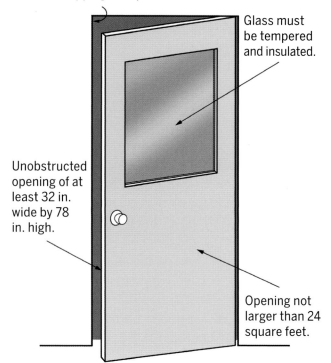

Glass must be tempered and insulated.

Unobstructed opening of at least 32 in. wide by 78 in. high.

Opening not larger than 24 square feet.

For attached garages there are special fireproof requirements.

2 Materials

If we lived in a Soviet paradise, this chapter would be short. At your local lumber yard there would be one type of wood and one type of plywood available. At the hardware store, you would have the choice of one type of glue and one kind of finish. My advice would be "use them to make doors." But instead, access to the global market economy has given us more choice than we really know what to do with. And so this chapter is long.

My advice is from experience tempered with some technical knowledge to the end of getting you making doors. Of course woods, glues and finishes are far more complex than covered here, so I'll do my best to point to other in-depth resources that will refine what I have to offer. Just because I don't say anything about thermoplastics, metals or straw doesn't mean you can't make doors with them. It's just that I don't know how.

Solid Wood—Choosing a Species

There are many hundreds of species of wood, varying considerably in their looks and physical properties—but not all of them are appropriate for doors. Wood is not a uniform material: ironwood will perform very differently from balsa in a door (I wouldn't recommend either). Within a good species for doormaking, the run of the grain within a given tree has different physical properties. Then, how you work the boards also influences the stability and longevity of your doors.

Looks and cost might be foremost in your mind—doors have to match the home they're going into, and I'll never suggest that you should take out a second mortgage to pay for the materials. Also consider the wood's workability, stability, durability and even its toxicity before making a choice. These will contribute to the life of the door and your enjoyment making it.

Workability is mostly a measure of the softness of the wood. Softer woods, such as butternut, are easier to cut but dent more easily, don't hold screws as well and can be more difficult to sand smooth and flat.

Stability depends more on the grain of an individual board than the species. Quartersawn heartwood is generally more stable than flatsawn boards and boards with pith or sapwood. That said, the mahoganies are generally more stable and maple is less.

Durability is a matter of the wood's ability to resist the effects of the elements and insects. Non-durable woods, such as maple and pine, may last only a year or two outside before they start to rot. Others such as black locust are naturally resistant to rot.

A wood with all the right characteristics—and that looks the way you want it to—can be hard to find. On the following pages is a list of some of the woods I'm most familiar with, their characteristics and my recommendations. Of course, the recommendations are not black-and-white: I know of exterior doors made in pine that have lasted 200 years through careful use and upkeep.

If you don't find the wood you're interested in, there are more complete resources on wood characteristics, such as *World Woods in Color*, William Lincoln (Macmillan/Linden 1986, etc.) and of course the Internet—the sum of all our unedited prejudices with some information mixed in.

Alder (alnus rubra) (33 lb/ft3). Not Recommended. Alder is mostly a West Coast wood, but lately is easier to find across the US. Great for small, non-structural work, it's a pleasant reddish brown and fine-grained in appearance, inexpensive, abundant (and therefore "sustainable"), soft and weak, and not particularly stable or durable.

Ash (fraxinus americana) (41 lb/ft3). Recommended for interior. Mostly known as the best wood for a baseball bat, white ash is commonly available. It is a nearly white, coarse-grained wood, hard and stable, but not durable against weather and insects. The Emerald Ash borer is currently decimating the ash population, but it is still available and inexpensive.

Beech (fagus grandifola) (46 lb/ft3). Recommended for interior. Beech is generally overlooked and not commonly available. A light tan wood, it is neither iridescent like maple, nor inexpensive as birch. Its density and fine grain make it a common choice for workbenches.

Birch (betula papyriferia) (39 lb/ft3). Recommended for interior. Paper Birch is often thought of as the poor cousin to maple, as they look alike but birch is cheaper and doesn't have the same luster. But it's a great workmanlike, fine-grained wood, easy to work and it's pretty stable. It is not durable against weather and insects.

Black Locust (robinia pseudoacacia) (34–54 lb/ft3). Recommended for interior and exterior. Black locust is one of the hardest woods I've worked with, durable like nothing else, and enormously heavy. You'll find it's tough on tools. Freshly cut, it's a greenish coarse-grained wood, but it turns a delicate brown with an oil finish and time. It's also hard to find, as most consider it better for tomato stakes and fence posts.

Butternut (juglans cinerea) (28 lb/ft3). Not recommended, but that breaks my heart. Butternut is beautiful and a favorite fine-grained wood, but sadly is lightweight, soft and has poor compression. Joinery in butternut will not take much stress before compressing and failing. Not a good wood for doors, except decoratively; but if you make a door out of it, I'd understand completely. I have.

Cedar, Western Red (thuja plicata) (24 lb/ft3). Sort of recommended for exterior. Western red cedar is weather resistant and durable, not too expensive and quite pretty. The trouble is it's extremely lightweight, scratches and dents easily, and so acidic that it will eat steel screws and iron hardware for lunch. So use it for light duty exterior doors, stick with brass or non-ferrous metals for hardware, and the doors should come out fine. Though it does have a spicy odor, it is not the same wood as eastern or aromatic cedar (juniperus virginiana) which is highly stinky and great for lining closets where you don't want bugs.

Cherry (prunus serotina) (36 lb/ft3). Recommended for interior. Cherry has become a premier fine-grained cabinet wood for its deep luster and dark reddish hue. It has a Goldilocks weight and density (not too heavy and hard, not too light and soft), making it easy to work. It is, however, a &*#^% pain to finish, as it tends to blotch, even with an oil. Cherry is moderately durable against weather and insects, but it's still primarily an interior wood. It also darkens considerably in sunlight.

Douglas Fir (pseudotsuga menziesii) (33 lb/ft3). Recommended for interior. Among softwoods, Douglas fir is a better choice than pine, and even widely available in inexpensive 2 x 8 stock, but it still suffers from the problems of all softwoods. It's lightweight, dents and crushes easily. Fir is more durable than pine, but not to the point of being a great exterior wood; hard to find except as 2-by construction material.

Pine (pinus radiata) (30 lb/ft3). Recommended for interior. Pine can make a fine door, but you'll have to overcome a few major hurdles. Pine doesn't take a finish particularly well, and it's not durable. It dents and crushes easily, so pegging your mortise-and-tenon joints won't work well (wedge them instead). Also, most home centers and lumberyards carry only ¾ select pine, so finding thicker boards for doors can be difficult. On the other hand, pine is supremely easy to work and inexpensive. I use pine for rustic projects. For paint grade work, use soft maple or poplar.

Mahogany (swietenia macrophylla) (34-40 lb/ft3). Recommended for interior and exterior. It grows all over Central and South America, and so has many names, including Honduran, big leaf, genuine, and South American. It varies considerably in density, color, and the quality of the grain (fine), but the better boards are a beautiful reddish brown, the wood a delight to work with, really stable, and insect and weather resistant. However, it is becoming harder to find and more expensive due to international restrictions on its exportation.

African Mahogany or Khaya (khaya senegalensis) (34-36 lb/ft3). Recommended for interior and exterior. Offered as a substitute for genuine mahogany, African mahogany looks similar but is not as easy to work. Durable, stable and fine-grained, it should serve you well, though it has issues with cracking across the grain and reacts with iron creating black stains.

Poplar (liriodendron tulipfera) (31 lb/ft3). Sort of recommended for interior. Poplar is an inexpensive 'paint grade' fine-grained hardwood that can end up looking quite pretty with a clear finish. The sapwood is white and the heartwood is greenish with black streaks that age to dark tan. It's among the less dense hardwoods, easy to work, relatively stable if not particularly strong. But for low-use interior doors it's perfectly good.

Meranti (Shorea spp.) (41 lb/ft3). Sort of recommended for interior and exterior. The meranti lumber sold in the US is from the same family as lauan, but it's not the same stuff as the super cheap and soft plywood. Some consider it a mahogany substitute, but it's inferior in nearly every respect, from strength and durability to stability and quality of finish. It is durable and beautiful, and can be used successfully in laminated constructions, but I wouldn't recommend solid frame members made of it.

Soft Maple (acer rubrum) (39lb/ft3). Sort of recommended for interior. Soft maple has many names, including red, swamp and silver. It is an inexpensive fine-grained wood that looks somewhat like hard maple, but has darker heartwood, often with mineral streaks. It's easy to work and somewhat stable.

Hard Maple (acer saccharum) (45 lb/ft3). Recommended for interior. Hard maple sometimes goes by the name sugar maple. It is hard, heavy, lustrous, light and fine-grained, but difficult to work and not particularly stable. This means you really have to choose boards well and spend extra time and effort milling to ensure your door members are straight and stay that way. It is a terrible exterior wood, as insects just love to eat it.

Red Oak (quercus rubra) (48 lb/ft3). Recommended for interior. Red oak is a light, coarse-grained wood that's only red in comparison with white oak (which is really tan). It's strong, stable, hard to work, heavy and reasonably inexpensive. There are actually many subspecies of red oak, so color, grain and character varies considerably. High in tannins, using iron hardware will lead to black stains.

White Oak (quercus alba) (47 lb/ft3). Recommended for interior and exterior. White oak is only called white because it isn't red oak. It has a toasted tan look, is heavy and hard to work, but is strong, stable, coarse-grained and ages well. It is durable against the elements and insects, so you can use it for exterior doors. There are also many subspecies of white oak, so color, grain and character varies considerably. In its quartersawn figure, it is much more stable, but due to the ray fleck, can look like snakeskin. Rift sawn oak is more stable than flatsawn and without snakeskin. As it is high in tannins, using iron hardware will lead to black stains.

Spanish Cedar (cedrela odorata) (25–30 lb/ft3). Not recommended. Spanish cedar looks somewhat like mahogany, is weather resistant and is inexpensive. On the other hand, it is lightweight, soft and has a strong, spicy smell. Its best use is cigar boxes, not doors.

Pear (pyrus communis) (44 lb/ft3). Not recommended. Pear is one of my most favorite woods, but largely for its looks. The color ranges from a pinkish tan with the warm depth of cherry to a whitish tan with little character. Though heavy and fine-grained, it is a fairly weak wood subject to brash failure (cracking across the grain). It bleaches badly in the sun and isn't durable, so don't put it near a window. It is also expensive.

Walnut (juglans nigra) (40 lb/ft3). Recommended for interior and exterior, with qualifications. Black or American walnut is a beautiful fine-grained wood that's a delight to work with. It's not too hard, quite stable and durable. I've made many doors in it, even some exterior doors, though it bleaches in sunlight and so needs overhead protection. Also much walnut is steamed so that the white (and non-durable) sapwood turns the color of the heartwood. And finding straight and clear walnut without knots is hard. Sadly, it is becoming more popular and thus more expensive.

* Weights given are averages, and may vary considerably from tree to tree. They're also calculated at 12% moisture content.

** One cubic foot is equal to 12 board feet. So a wood with an average weight of 36 lb/ft3, should come out about 3 lbs for every board foot.

***Recommendations are exactly that, and not strict rules. I've made exterior doors in walnut and interior doors in pear, and they're lovely in spite of their problems.

NOTE: Weights/density figures are from *World Woods in Color*, William Lincoln (Macmillan, Linden 1986)

Wood and Health

While the Occupational Safety & Health Administration (OSHA) won't come out directly and say wood dust causes cancer in addition to other ailments, they do say they are "associated," so take precautions with dust masks or respirators—as an old friend used to say, it's just not wise to use your lungs as the dust filter for the shop.

Indeed, when I first started out 30 years ago, only some of the exotic woods bothered my nose. Now, I use respirator because even plain old pine dust leaves me coughing and sneezing by the end of the day.

LUMBER TYPES IN CROSS-SECTION

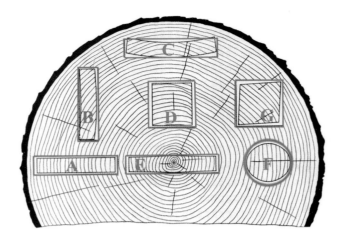

In cross-section, A through E represent commercially available lumber types and how they move when they dry. A is quartersaw, B and G are riftsawn, C and D are flatsawn. While E may be quartersawn, it is not stable and will warp and crack due to the presence of the pith (the heart of the tree). Avoid the pith in making doors. (Forest Products Laboratory, USDA Forest Service, 1999. *Wood Handbook*, Wood as an Engineering Material, General Technical Report, FPL-GTR-113, US Department of Agriculture, Forest Service, Forest Products Laboratory, Madison WI. p. 3–8)

Good stewardship

Trees are a renewable natural resource—I don't feel badly about cutting down trees in my town and turning them into furniture because the forests in my home state of Connecticut are growing. This is, however, not the case everywhere in the world.

Some very beautiful lumber comes to US markets illegally and from countries that are happy to take it from forests in an unsustainable manner. Some species have virtually disappeared from the market due to overcutting, notably the legendary Cuban mahogany (swietenia mahogoni). I believe in good stewardship—being thoughtful about what we use and how we use it—so I don't like contributing to this trade.

At the same time, there is a lot of exotic lumber that is taken from forests in a responsible manner so I do not forswear lumber that doesn't come from Connecticut. At the risk of being awkward, ask your dealers about the sources of their lumber. Ultimately they may not know, but good stewardship begins with small steps— They might be inspired to find out.

Wood Moves—Deal With It

All wood moves. After you cut it to size and shape, it will change size and shape, continually and cyclically, forever more. Some of this movement is not controllable, but much of it is manageable when you know why it moves and when it moves. It is key to minimize wood movement in a door, as retaining its basic flat, rectangular shape is key to its function. Doors that have turned into potato chips don't close or open easily, keep out the bugs or much else.

Wood moves for two basic reasons. The first is the internal stress built into the grain during the tree's growth. Trees that grew with curves or twists in their trunks or branches will have the most internal stress, as will wood around knots. This will be expressed in the lumber after it's sawn straight as twists, kinks, bows and crooks. It may be possible to mill these boards flat again, but they will always have the same tendencies and so are best to avoid when you need stability. The straighter the trunk and fewer the branches, the less stress there will be to warp boards. Additionally, annular growth rings will tend to flatten out with time, so all flat-sawn lumber has a tendency to cup. This type of movement you can minimize greatly in your choice of grain orientation and board characteristics.

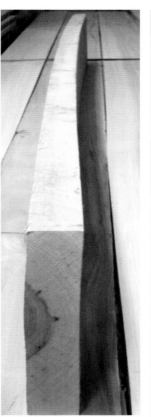

Above: This breadboard end was cut flush with the table edges in August, but now it's November: proof of how much wood (soft maple in this case) can move.

Right: Bowed boards will continue to bow after milling.

Far right: Crooked boards will continue to crook after milling.

The second reason wood moves is moisture content. As the relative humidity in the air around wood increases, wood will grow dimensionally. When relative humidity decreases, wood will shrink. This movement is mostly across the thickness and width, not along the length (which hardly ever changes). This movement can be minimized, though to a lesser extent, by specie choice (some move more than others) and the application of finishes that retard the absorption and loss of moisture.

How much movement to expect from these sources is difficult to predict with any certainty (though there are many books with species-specific information to help). On average, I've come to expect that most North American hardwoods will grow and shrink about ⅛ in. to ³⁄₁₆ in.—even ¼ in.—across 1 ft. of width in a flatsawn board between the wettest and driest times of the year. A three foot wide table top can lose up to ¾ in. width in winter—this amount seems impossible, until you see it happen.

Here are some strategies to minimize wood movement in the wood you use for your doors:

Use straight-grained wood without defects

At least in my neck of the woods, most hardwood lumberyards are perfectly happy to let customers pick through the piles of lumber to find the specific boards you want, as long as you restack neatly when done. Of course, ask first.

Looking at a rough board will tell you what internal stresses it has. Twisted boards will always tend to twist more over time, bowed boards will always tend to bow more, kinked boards will continue to kink, etc. As you work the wood, flattening and shaping it, only some of these stresses are released. The rest will come out in time, eventually warping your door.

Straight-grained wood is most important in frame-and-panel designs, where the relatively thin rails and stiles comprise the entire structure of the door. Board-and-batten construction can accommodate more

Right: Kinked boards with grain run out will never be stable.

Far right: Cupped boards will continue to cup after milling.

Below: End checks are common in most boards, a consequence of uneven drying.

Below, bottom: Bend a thin cut from the end of a board to reveal hidden cracks.

imperfections since the larger boards can minimize the influence of any particular knot—but larger boards will also move more.

Cracks at the ends of boards

Most dried lumber has stress cracks at the end of each board, visible and invisible. They come from the drying process. It's important to get rid of them completely as they will tend to propagate in the finished door, weakening any joinery around them.

To avoid cracks in your finished door, first, buy longer lumber than you'll need. For an 80 in. tall door, buy at least 8 ft. long lumber to give yourself some room to trim the ends. The visible cracks are easy to avoid—cut them away where you see them. Many cracks, however, are not visible. To find them, you have to use your hands, not your eye: trim ¼ in. sections off the ends of your boards and try to break the grain across the width. If you can break it easily, you probably have a crack there. Keep trimming off ¼ in. sections and keep trying to break it until you have uniform resistance across the entire section. Then your board has no more hidden cracks at the ends.

In frame-and-panel construction, it also helps to cut stiles longer than necessary and trim them when the door is complete. It's also a good idea to seal the exposed end grain in the stiles, especially if it's an exterior door.

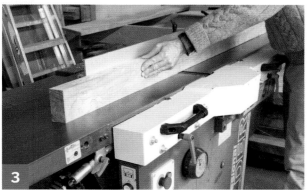

A thin coat of epoxy or an extra few coats of polyurethane will help prevent future cracks by slowing down the loss and absorption of moisture, especially on the bottom of the door.

Mill your lumber in steps

Who has not ripped a straight board to watch the two halves bow during the cut? Milling releases stress held inside boards, creating warps where there may not have been any, or accentuating preexisting ones. This is because the stresses within every set of fibers hold the others in check. Remove one set, and you allow the others to push the wood the way they want to. Sawing to width, jointing and planing will all reveal a board's tendency to warp, if any. This can be the most frustrating part of making doors, when you joint and plane a board flat and true, only to find by next morning that it has warped back into its original shape.

Boards that you need to be particularly straight, flat and true should be milled in stages. I think of it as releasing stress slowly so you remove the right amount of wood in the right places, almost to warp the wood flat. If you start with relatively stable lumber, this milling-in-steps process will give you stable, flat and true door parts.

Rough milling

This process is useful for every project later in the book that involves long, solid wood stiles. Start with boards at least 4 in. longer, ½ in. to ¾ in. wider and ¼ in. thicker than the finished dimensions.

1. Find and remove any cracks at the ends, as described above. I use a circular saw, but any handsaw or chopsaw will work.
2. Joint the faces flat.
3. Joint an edge perpendicular to the face on each board.
4. Rip the boards ½ in. to ¾ in. wider than finished dimensions, or oversize enough to accommodate any warping that happens after you rip them. For a board that's straight and in a species that doesn't move much, you can rip

them at ¼ in. wider. But for an unstable wood in a rough board that's already warped, I might rip ¾ in. wider.

5. Plane the opposite face to ⅛ in. to 3/16 in. thicker than finished dimensions. It is not necessary to clean up all the sawmill marks on the boards at this point.

6. Rest the boards on edge on stickers to allow air circulation all around. Leave them overnight.

Fine-tune milling

7. Check the boards in the morning to see if they have warped at all. Scribble pencil marks on all surfaces of the boards (to keep track of which surfaces you've milled and not).

8. If the board has warped, repeat the rough milling process taking off a minimal amount to flatten and straighten all four faces.

9. Rest the boards on edge on stickers to allow air circulation all around. Leave them overnight.

10. If the boards have again warped overnight, you'll need to repeat this process until they stop moving, or live with the warp. In theory, the less material you take off the less the board will warp. If a board keeps warping overnight after you've jointed and planed it flat several times, then you shouldn't use it for a door since it will continue to warp that way.

11. Plane the board to final thickness and rip the board to final width on the tablesaw.

What's good in sheet goods?

Plywood came into woodworkers' lives in the 19th century, a watershed moment that should be celebrated annually with bells and whistles. By laminating thin sheets of wood in perpendicular layers, you get dimensionally stable sheets that are also reasonably strong in every grain direction, largely canceling two of the major structural drawbacks of solid wood. What's not to like?

And yet, the use of plywood and other sheet goods for traditional doormaking are limited because they have weaknesses that solid wood does not:

- Not as much longitudinal strength as solid wood.
- The edges reveal the core materials (rails, stiles and raised panels will reveal this edge).
- The surface veneer can be extraordinarily thin (⅛th of an inch) making it easy to sand through and not durable.
- The surface veneers may delaminate over time.
- High-quality sheet goods are expensive (exterior and marine grade plywood is extortionately expensive).
- The adhesives used in sheet good manufacture can be quite toxic.
- Nevertheless, sheet goods have an important role in modern door construction. They work well for flat captured panels, as the edges do not show. And plywood is the essential ingredient for hollow core doors.

The right goods for the job

Choosing the right sheet goods can be confusing, as the options are many.

Commonly, plywood is sold in 4-ft. by 8-ft. sheets, a standard dimension for building construction. You can get larger and smaller sheets, but they're specialty items. And yes, if you're asking, the full-size pickup truck bed is sized to accept a standard sheet.

"Cabinet grade" plywood offers a high number of internal plies (generally 5-9), a smooth, void-free, sandable hardwood veneer surface that will take a high-quality finish. Construction-grade plywoods generally offer fewer plies (5 or even 3), voids and thick softwood veneer surfaces. I don't recommend using the latter unless you want that look and performance.

Your choice of surface veneer doesn't have to be based on the physical properties of that wood, as the structural properties of the sheet goods depends almost entirely on the core material. A sheet of walnut plywood will behave the same as a sheet of maple. However, a butternut veneer will still scratch more easily than a maple veneer.

Cabinet grade surface veneers come in grades and styles relating entirely to their appearance. Face veneers have letter grades and backer veneers have number grades, usually equal or inferior to the surface veneer. For example, cabinet grade plywood comes in grades of A-1, B-2 and C-4. No, there is nothing explosive about a sheet of C-4 plywood.

While grade parameters differ between species, the A's will tend to have fewer defects than the B's and so on. Exceptions are abundant—for example, A-grade maple can include black mineral streaks, not considered a defect by the manufacturers, though almost always by customers. On the other hand, the B grades generally don't have those mineral streaks and are sometimes sold as a "White B".

You also have choices of the type of grain (e.g. rotary or sliced, flatsawn, quartersawn)

Plywood vs. Sheet Goods

Since plywood's invention, core material options have evolved beyond "plies", or thin sheets of solid wood (not "piles": that's a medical condition). They now include a wide range of manufactured materials such as medium density fiberboard (MDF) and oriented strand board (OSB), each offering different structural properties. The more encompassing term now is "sheet good," though this makes me think of cookies out of the oven. Perhaps "veneer sandwich" has a more accurate and appetizing ring.

Sheet goods from top to bottom: Exterior construction-grade, cabinet-grade maple, Russian birch, interior AC grade, exterior 3-ply, MDF and marine-grade okoume. Note the different smoothnesses of the surfaces.

and the type of arrangement on the sheet (e.g. book matched, slip-matched, random).

Core materials

If that wasn't enough, then you come to the core materials which determine the structural properties of the sheet:

Particleboard core looks like old breakfast cereal, a coarse mash of fibers. Compared to solid wood, it is inexpensive, not durable and not strong. Standard screws will not hold in it well. But for a panel inside a frame that has no stress on it, it's perfect.

Medium Density Fiberboard (MDF) cores are just as weak and inexpensive as particleboard, but it's much heavier and denser. MDF's advantage is its flatness—panels will look more "perfect" with an MDF core.

Veneer and lumber cores are layers of solid wood plies. Structurally they are stronger and act more like solid wood than the manufactured materials. This means they move, though in odd ways (see next section), and hold a screw reasonably well (though not into the edges so well).

Baltic or Russian birch is an entirely different sheet good product than standard cabinet grade plywood. It has thinner plies offering a more even look on the edge, though no structural advantages. The higher grades are "void free," so the edges look even and can be exposed without looking too shabby. It more commonly comes in 5-ft. by 5-ft. sheets, but is also available in 4-ft. by 8-ft. sheets. For a captured panel, Baltic birch has no advantage, though it can be used to make a solid plywood door with nicer edges.

Marine-grade plywoods are just that—used for boat building. These plywoods are waterproof, have no voids in the plies and are guaranteed to be structurally sound. While I'm not a boat maker, I have used marine grade plywood for exterior applications with good success, but still don't recommend it.

For flat panels in interior doors, I have no problem using veneer core plywood. I stay away from the MDF and particleboard cores as the advantages of a slightly lower cost and better flatness don't outweigh the loss of strength and addition of weight. I also don't use them in hollow core door construction for the same reasons. Call me a belt-and-suspenders purist. I've been called worse.

Plywood adhesives

Standard sheets of plywood use ureaformaldehyde resin glues. They are cheap, waterproof and really durable—except they outgas formaldehyde. How bad is it for you? Well, I just don't know, but the government considers it a carcinogen (just like wood dust), so I take precautions. Just as I use a respirator, I keep my shop well-ventilated when working with plywood. Some manufacturers make plywoods that use more environmentally-friendly adhesives, but at a greater cost. If you have more concerns or questions, the Consumer Product Safety Commission (www.cpsc.gov) offers a Publication (725) on formaldehyde.

Plywood doesn't move—
eppur si muove

After recanting his belief that the earth moved around the sun, Galileo is said to have muttered under his breath, "eppur si muove"—and yet it moves. Such is the truth of plywood. It doesn't move. And yet it does—just not in intuitive ways.

The veneers in plywood do cancel out their cross-grain expansion and contraction. It is dimensionally stable. But plywood still breathes the surrounding air, taking in and releasing moisture with changes in relative humidity. This creates and releases internal stresses, making veneer core and lumber core plywood warp through twisting and bulging. Stand a full 4-ft. by 8-ft. sheet of ¾ in. thick plywood on edge, and 99 out of 100 will have a bow, a cup, a twist or a combination

of the three. Move it to a damper or dryer room, and over time it will flatten out then warp in the other direction. Lean it against a wall so there's less air circulation on the back side and it will change shape again. Unlike solid wood, small warps cannot be machined out of the sheet. Thinner plywood is even more prone to warping.

The only way to cancel this warping is to attach the plywood to something else. The construction of a hollow-core door does just this, enabling you to make a flat and stable door from warped pieces of thin plywood. Still, that warping stress will always be within the door.

MDF and particleboard core sheet goods do not warp in the same way, as they do not have an aligned grain structure. They do absorb and release moisture, however. If one side gets more moisture than another, they will warp (and begin to disintegrate).

Using plywood or veneer on exterior doors, I have found, does not always work too well. If exposed to direct sunlight and rain, the extreme expansion and contraction of the exterior surface stresses the adhesive between veneer and substrate,

How Flat is Plywood?

A good friend once bought an expensive, extremely accurate 36 in. wide thickness sander figuring it was a great investment in reducing the amount of time he spent sanding by hand. And it was: a few passes would eliminate planer marks and tear out, and leaving a much flatter surface compared to hand sanding, and it took a fraction of the time. Then he got thinking about the exotic veneered plywood panels he used in his cabinetry—the surface veneer was thin, but using 220 grit and a light pass should not cut through the veneer. And how much sanding time would that save? Tons, of course!

The 24 in.-wide test piece came out the other side and he laughed: it looked like a topographical

map, with the veneer sanded through completely to the core material in round and oblong peaks here and there, while other spots hadn't been touched.

And so he learned the hard way that veneer and lumber core plywoods are surprisingly not-very-flat and are uneven in their advertised thickness. Sometimes the plies accidentally overlap. Sometimes a ply gets left out. I've measured a sheet that was ¾ in. at one end and ⅝ in. thick at the other.

The thickness of MDF and particleboard core panels are much more even, but I have not sent a test piece through my friend's thickness sander to see just how even.

eventually leads to delamination even using exterior-grade plywoods that have good protective finishes. I keep my life simpler by sticking with solid wood.

Glass and Screening

While I believe in a do-it-yourself ethic, there are some materials that I could conceivably make myself, but don't, and don't recommend trying to make: these include glass and screening.

By code, glass panes set in doors must be tempered. For exterior applications they should be insulated (double or triple pane) to attain the R-value required in your area. Other than that, you have your choice from the many different types of glass available. The most common type is flat, nearly colorless float glass. It's inexpensive in ⅛ in. thickness and perfect for interior French doors. Glass can be frosted or textured, have tints and UV blockers, even colors if you like. All of these can be cut to any size and shape.

If you're working with non-rectangular glass panes, I recommend making two identical templates of the shape out of cardboard or thin MDF. One template goes to the glass shop and the other you keep to ensure they've delivered the right shape (I have never had a template returned to me—the explanation is that they get "damaged or lost."). Local glass shops are always glad to help winnow the choices to your application and budget.

Screening comes in a wide variety of materials with a range of properties. Fiberglass is the most commonly available material. It is a bit stretchy and doesn't crease or dent making it easy to install. It doesn't rust, but it does tear easily. It is also inexpensive. Vinyl coated polyester mesh is much more durable and a good option if you have pets or small children. Metal screen meshes are more durable, but they easily kink or crease and corrode over time. Aluminum is the least expensive and most common. Bronze, copper, brass and stainless steel are far more expensive but the ways in which they corrode look good (i.e. they gain a patina.) Metal screening, however, is best used in a like-metal frame or in a non-corrosive wood.

Whichever material you use, I recommend darker colors as they are easier to see—helping you avoid walking through them.

Adhesives

Choosing a wood glue is rarely an issue of strength, no matter what advertisers say. All modern adhesives, even Elmer's white glue, form bonds stronger than the lignin that holds wood itself together—that is if you apply the glue according to its specifications and thereby get a good bond. Glue line failure happens mostly due to some overlooked or misunderstood factor in the application. I've been there, hopped up and down mad, wondered what the heck went wrong, and finally sat own and read the instructions. And using glues effectively begins with following the directions on the back of the bottle.

Factors that make-or-break adhesive bonds include curing temperature, the amount of open time you take, the wood surface texture, how perfectly the mating surfaces come together, the grain orientation, the wood's moisture content, the type of wood and its resin content, how long before the joint was cut, the clamping PSI, the drying time, the age of the glue and how it was stored and if the gremlins in your shop are in a good mood. Only when you get everything within the range of what a particular adhesive requires, can you depend on the bond really being long-lasting and stronger than the wood itself. And that range is different for every type of glue.

There are no particular glues for making wood doors other than wood glues. Exterior doors do beg for waterproof glues while interior doors do not. My choices are based more on ease of use and the particular application— whether I need a long open time for a complex assembly, a quick tack because clamping is difficult, or gap-filling because the joint is loose.

Adhesives are like finishes in that they have a learning curve to gain the technical knowledge and skill necessary to use them well. I highly recommend using the glues that you know well, and testing new glues before using them on a door. The one good glue book I know is William Tandy Young's *The Glue Book* but it is close to 20 years old. I am no chemist, so the information and recommendations that I give below are in layman's terms and might run roughshod over the precise details.

White and yellow glues

The glues I use most often are polyvinyl acetate (PVA), commonly known as white and yellow glues. They are inexpensive, clean up with water and are reasonably non-toxic (though I wouldn't recommend eating the stuff).

On average, they have a fast tack, which is to say they will form a permanent set relatively quickly so you have to be fast with a clamp; however, manufacturers do offer a wide variety of PVAs with a range of properties from quick to slow-tack, water soluble to waterproof and thin to thick.

PVAs are also thermoplastic: they remain relatively soft and will turn liquid with high heat. This makes them a poor choice for glue joints that remain under pressure, such as bent laminations, as the glue joints will creep, or slowly side apart.

For a good bond, the mating surfaces need to be smooth and flat. If the gap between them is greater than $\frac{1}{1000}$th of an inch, the bond gets weak. PVA's do not "fill gaps," so if the joint is loose, PVA will not work anywhere near its full strength.

Polyurethane

Polyurethane glues are water resistant, have a slow set and virtually no tack. While they're more expensive than PVAs, you don't need as much on the joint surface as they expand while curing. However, they are not gap filling and require a solvent to clean them up. If you don't want to wash your hands in solvent, wear gloves.

To bond well, the glue surfaces should be smooth and flat, though the tolerances are wider than for PVAs. Polyurethane needs moisture to cure, so when gluing kiln-dried woods, first dampen the joint surfaces with a sponge. I find that the main advantage of polyurethane glue is its low initial tack and slow set. The joints in a large project will slide together more easily with wet polyurethane glue on the joint surfaces, making adjustments easier. And you will have more time to adjust before the glue sets.

Ureaformaldehyde

Two-part ureaformaldehyde based glues make strong, waterproof, and UV-resistant bonds with a slow set up time. And they're quite cheap. What's not to like? Well, you have to mix the parts, it doesn't fill gaps, doesn't cure well in low temperatures and has formaldehyde in it. The latter is why they rarely write "UREAFORMALDEHYDE" on the label, but are instead marketed with names such as Resorcinol, Unibond, Pro Bond, Weldwood or Plastic Resin glue. While I used them all the time when I was an apprentice, other glues such as water resistant PVA have become more popular since they perform well enough and do not require mixing. Still, because of their slow set up time, they're perfect for large laminations. Use in a well-ventilated shop.

Structural epoxies

First of all, don't confuse the "5 minute epoxy" in little tubes available in most hardware stores with real structural epoxies. I'm not sure what those glues are good for, but I haven't found it yet.

True structural epoxies, such as West System and System 3, are primarily used in boat building and building restoration. These are truly waterproof and gap-filling adhesives (and until you can cut perfect joinery every time, gap-filling is a word of grace). You can adjust their set time and thickness by using different hardeners and different fillers. On the down side, they are expensive, require a solvent to clean, and allergenic. While I don't have a problem with epoxy, some friends develop a bad rash from touching it. So use gloves and goggles at a minimum.

Epoxy's gap-filling properties are its primary attraction. This means that the adhesive itself has structural strength, so it doesn't matter how close or far the mating surfaces are (though there has to be room between them for the epoxy). Epoxies form stronger bonds when the mating surfaces have some "tooth" in them, a roughness that gives the adhesive better purchase, and a bit of a gap, so you can't clamp the joint as tightly as you would for a PVA or ureaformaldehyde glue.

Everything else

Cyanoacrylate glues, often called "super glues," have a near-instant tack making them perfect for small veneer patches that are hard to clamp. But that near-instant tack makes them somewhat hard to work with in large joinery. Hide glue, the standard (only) glue of much of Western furnituremaking history, is still beloved by restorers because it's reversible. A friend swears by urethane glues and has tried to get me interested in them. But as he works with metal and glass so I ignore him. Construction adhesives are gap filling and contact cements can be nearly instantaneous. Yet I've not found much use for these glues in doormaking.

Finishes—Protection and Beauty

You can leave your completed doors unfinished if you like. They'll hold together for quite a long time—indefinitely if they're interior doors. In fact, if you're an amateur woodworker who has had a few of those finishing disasters that come with the learning process, you may be sorely tempted to do just that with your nice new door.

There should be no mis[...] finishing is just as complex [...] discipline as woodworking [...] opening a can that lists as s[...] on the outside and applying [...] bit like playing Powerball—[...] of winning, but it's slim. Th[...] is to learn how to finish (or [...] finisher). For DIY finishing [...] Flexner's *Understanding Wo*[...] good resource even though it's a few years old.

Modern finishes are changing rapidly, largely due to new EPA rules on emissions. In short, waterborne technologies are getting better and solvent-based finishes are getting worse. The spar varnish that you could buy in 1970 would most certainly outperform the spar varnish you can buy today because the formulas have changed to be less toxic to you and the environment. The professional finisher I work with often complains that good general advice is hard to give because specific products are changing so fast, new ones are being added and long-term performance is hard to gauge since many products haven't been around that long.

Now, I make no claim to being a finishing expert (talk to Bob), but I'll share some basic finishing tips for doors from my experience. The main one is that exterior doors really benefit from a high-quality finish to keep the look you like, slow down degradation and avoid rot. Otherwise, finishing doors is no different from finishing furniture.

Interior door considerations

The main consideration for interior doors is the ease of cleaning. Dirt tends to accumulate around the handle while the lower rail tends to collect scuffs from shoes. If the door has frames and panels, then, just like a picture frame, the lower horizontal moldings tend to pick up dust.

Film finishes (everything except the oils) and most paints are easier to clean—a damp rag gets the job done in seconds. Penetrating oil finishes make cleaning a little harder, sometimes allowing the dirt or scuff to get into the wood. To remove dirt, you'll need to use steel wool or sandpaper. For a rustic look, however, this "patina" can be part of the charm.

If you have a favorite finish you have used satisfactorily for table tops—i.e. a surface that gets a fair amount of contact with hands—it should work well for interior doors.

Exterior door considerations

If you like that gray weathered look, don't apply a finish and you'll get one pretty quick by leaving the door in the sun for a day or two. It's UV light that turns all wood gray, but only on the surface—the effects don't go deep (this is why oxalic acid—deck brightener—or the lightest of sanding reveals the original wood). So there's no harm done except you've lost the look of the wood. However, at the same time the sun is turning your door gray, it's also heating the surface substantially, shrinking the surface wood and potentially cracking it or opening the joints. These cracks provide entry for water, both atmospheric humidity (interior doors) and rain (exterior doors), that does serious damage, degrading and eventually rotting the door. A gray, weathered door will only last indefinitely as long as you keep it dry. I think.

Finishes act as a barrier to sun and water, preserving the wood from the effects of both. The most effective finishes to protect against the sun are paints—pigments effectively block UV light. But if you've gone to the trouble of making a door in a beautiful hardwood, you might want a clear finish to see that wood.

In a frame-and-panel door, the seams between the panels and frame offer moisture a way in that can hasten rot. Solid wood panels expand and contract, so there is no way to completely seal the gap between panel and frame. Using a rot-resistant wood helps, as does finishing the panel and inside grooves of the frame before assembly.

Solid wood endgrain in contact with the ground wicks moisture up into the joint, rotting it from within.

Another major route for moisture to get into a solid wood door is through the end grain of the bottom of the stiles. To help avoid rot, I seal the bottom edge of the door with epoxy before installation. There is no need to apply finish over the epoxy.

What you're looking for in a good exterior clear door finish is durability, flexibility and UV inhibitors. A durable finish will not wear away. A flexible one will not chip. And the UV inhibitors will preserve the original color of the wood.

No matter how good the finish, the more sun and rain the door gets (from a Southern exposure if you live in the Northern hemisphere), the more often the finish will need to be renewed. UV inhibitors literally wear out and stop working before the finish wears away. The first signs of dullness in the finish and fading are announcements that the surface coats have degraded or disappeared and it's time to refinish.

Two-part urethanes: These industrial level finishes are perhaps the best finish for exterior doors. They are tough, durable and flexible, have UV blockers to preserve the color, and will even give you the smooth, beautiful clear look you associate with lacquers. That said, they are a highly technical finish that has to be mixed accurately, sprayed according to specific coat schedules, and so require above-average skills and equipment to apply. They are also difficult to repair if damaged. If I need this kind of finish on a door, I ask my finisher to do it.

Exterior varnishes: Advertised as "Door and Window" exterior finishes, they often won't say "varnish" on the can but will say "alkyd." While they aren't as durable as the two-part urethanes, they are flexible and can be applied by brush. They are also easy to touch up if there's a scratch. These are the finishes I generally use for doors as they mix performance with ease of application.

Acrylic urethanes: These waterbased finishes perform quite well, but are not as flexible as the exterior varnishes. They are also a cinch to use, as the cleanup is with soap and water.

Exterior polyurethanes: Polyurethanes are somewhat brittle and a much better interior finish. Exposed to direct sunlight, they may crack and chip away.

Exterior grade oils: Penetrating oils are delightful to use, and thereby favored. They will not chip or wear away, but they also do not protect the wood well. If you use one, plan on refinishing your door every year.

I won't recommend any particular brands—after all, they may change their formula next year. But I do recommend two other things. The first is a high quality brush. Most of the ease of application and the results you get comes from the quality of brush you use. And no I don't have any connection with Purdy brand brushes (though I do use them, just for the name). The second is to talk with a knowledgeable finisher or salesman about

Use a combination of power and hand sanding to produce scratch-free surfaces before sanding.

what finish you should choose specific to your local climate and other specific application.

Finishing preparation

The most important tool in the shop is your sander, and the most important technique is prepping surfaces for finishing.

Has this book author gone mad? Perhaps, but here's my logic: If I've learned one lesson over the years from my customers, it's that a good finish makes or breaks every project. The appeal of a solid wood door is in part its design, the choice of hardware, and its jamb and casing. But it's a great finish that highlights the wood, justifying the time and effort of using wood in the first place.

A great finish rests on the foundation of good surface preparation. Good surface prep comes through an attention to detail while you're sanding. A lot of woodworkers I know hate finishing because it seems so easy to go wrong (and golly, I've been there). The secret to a great finish is good surface preparation—with that foundation the finish will actually do what it says on the can.

In short, don't skip grits, use a particular grit until you can't see the scratches from the previous one, use low-angle slanting light to look for scratches if your eyes are as old as mine and if you can still see scratches, remember that the finish will likely highlight them, especially stains and paints.

Joinery Techniques and Tools

3

If you were paying attention earlier, you'd know by now that doors undergo some particular stresses. Keeping them together is the responsibility of the joinery you choose. Choose well, and build well, and you won't have to fix things later. Are you intimidated yet? Well, don't be. Good door joinery isn't that complicated, just sometimes time-consuming.

With the notable exception of chairs, doors undergo much more stress than most furniture. Doors are suspended from the floor by hinges along one side, leaving the weight of the lock side to pull down on everything. Tension, racking, shearing—you name it, doors have to resist it. The heavier the door, the stronger the joinery needed to keep everything together.

My rule of thumb (or prejudice) is to use strong mechanical joints, rather than glue joints, to hold doors together. Modern adhesives are wonderful—and are the only way to create several of the doors in the projects section—but I still find they are trickier to get right than pegs, mortises and tenons (perhaps if I just followed the instructions on the bottle....?).

By mechanical joints, I mean joinery in which solid wood resists the stresses trying to pull the joint apart. For example, a mortise and tenon is a mostly mechanical joint: it resists every stress put on it except tension (glue and/or a peg does that job). On the other hand, a biscuit joint is primarily a glue joint, as it needs glue to resist lateral shearing, tension and racking. It is strong only against shearing forces across the boards' thickness.

Failed joinery leads to sagging doors that don't close.

There are many hundreds of ways to connect two pieces of wood. This section covers the techniques I've found that work well holding doors together. It's not exhaustive, as that would require several books, but representative and recommended. In the projects section, you'll see how each can be used in a specific context.

Clinched Nails

Good doormaking techniques begin with the simplicity of clinched nails.

The shear strength of a common nail is pretty good. But nails do not do well under tension. In fact, a nail will work its way out of a board on its own due to the expansion and contraction of normal wood movement. Eric Sloane memorably describes how "softwood 'breathes' its nails outward," observing the way cedar shingle nails grow proud through the seasons.

To keep a nail from wiggling back out, the solution is to clinch it: use an over-long nail

CLINCHED NAILS

Nail driven through two boards and clinched back resists tension.

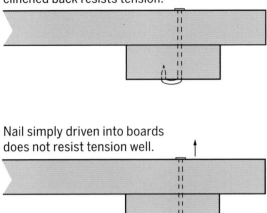

Nail simply driven into boards does not resist tension well.

THE PEGGED MORTISE-AND-TENON

All stresses, except tension, resisted by the large surface area of the joint faces. The peg resists tension mechanically.

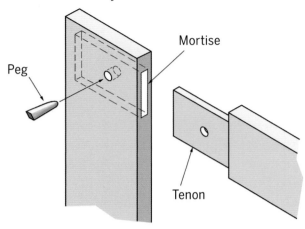

Peg

Mortise

Tenon

and bend the tip on the other side. While a single clinched nail can resist shearing, tension and twisting, it does not resist racking. Several clinched nails, however, do resist racking. This is why you will always see clusters of them on old doors. Working together, many clinched nails are a traditional and lasting way to connect boards without adhesives.

Bending a nail with a hammer is easy to do—we've all done it without trying, right? With clinching it's just a matter of bending it in the right places at the right time, on the other side of the board, after the nail goes through straight.

You can clinch pretty much any nail. Hardened nails don't really exist as they'd shatter when hit with a hammer. Wire nails (circular profile) are the most commonly available type and work perfectly well. Square or cut nails work well too, just look older and more rustic. To reduce the chances of splitting the wood as you nail, pre-drill small pilot holes through both members and clinch the end across the grain. For cut nails, align the chisel point across the grain, again to reduce the risk of splitting.

The Pegged Mortise-and-Tenon

In a small but significant leap of complexity, the other joint I recommend for doormaking is the pegged mortise-and-tenon. If made with some attention to detail, this joint resists every stress put on it quite well, including tension and racking. It can last several lifetimes of hard use. The key is that no stress is resisted by the adhesive alone, but by the surfaces of the joint. This can't be said of any other joint except the wedged mortise-and-tenon, and that joint is another leap in complexity.

This joint, however, either requires special tools or can be time-consuming to make. For this reason, many woodworkers look to faster and nearly-as-good options such as floating tenons (Festool's Domino system), Miller dowels, even simple biscuits and pocket screws. For smaller, lighter doors, these joinery systems can work well. But for a full-sized frame-and-panel door in solid wood, the pegged mortise-and-tenon is the gold standard for durability and longevity.

Both the mortise and the tenon parts of the joint can be made with simple hand tools, though this approach takes some practice and some patience. For mortising, the next step up includes the drill press and a full-size plunge router. But the cat's pajamas are the hollow chisel mortisers and slot mortisers, both dedicated machines that are not cheap. For tenoning, a dado blade or a tenoning jig for a tablesaw can work well, though industrial tenoners do exist. However there is no "right" way to make them: my advice is to use what tools you have, as you know how to use them, and they incur no extra cost to do so.

The mortise-and tenon joint without a peg, however, is not much better than a floating tenon joint. Without the peg, tension is mostly resisted by the adhesive, which, in my experience, tends to work loose over time. I think this is due to cross-grain movement in the relatively large surface area of door joints. Or I don't follow the instructions on the glue bottle. On the other hand, I've never had a pegged joint pull apart.

Square Pegs DO Fit in Round Holes

You can accuse me of having failed Kindergarten, but I believe that hammering square pegs into round holes works best. Square pegs don't come out as easily, as their corners cut and compress the wood as they're hammered in, helping secure them in place. Over the years, pegs do stand a little proud, the cycles of humidity levels expanding and contracting the wood. If they ever fall out, then the hole was too big or the peg too small to begin with. To fix the problem, hammer in a slightly larger peg.

The pegs should be made from a hard wood, as hard or harder than the wood in the door. On a cherry door, I might use cherry pegs to help the pegs blend in (though end grain is always darker than face grain). Otherwise, I often use white oak pegs no matter the species of the door. It's a hard wood and the end grain is interesting for anyone who gets up close to look.

Mortises can be cut quickly with a large plunge router, a long straight bit and a guide fence.

Tenons can be cut quickly using a crosscut fence and dado blade on a tablesaw. The rip fence can act as a width stop.

A square ¼ in. peg driven into a ¼ in. round hole through both mortise walls and tenon will lock the joint together.

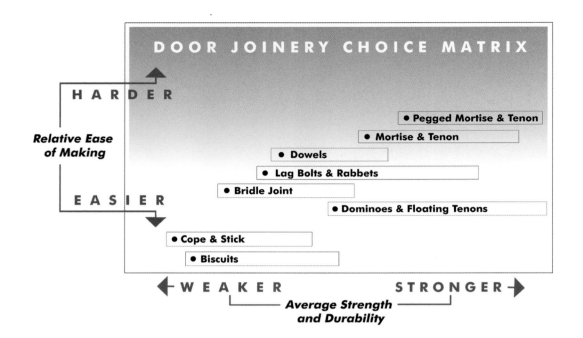

DOOR JOINERY CHOICE MATRIX

HARDER

Relative Ease of Making

EASIER

• Pegged Mortise & Tenon
• Mortise & Tenon
• Dowels
• Lag Bolts & Rabbets
• Bridle Joint
• Dominoes & Floating Tenons
• Cope & Stick
• Biscuits

◀ WEAKER   STRONGER ▶

Average Strength and Durability

Stave Construction

The most stable solid wood construction involves laminating several strips, or "staves," together in a sandwich. Each stave counters the stresses of the other, so the construction warps much less over the long term than solid wood. The addition of several glue lines also adds strength, as glue joints are stronger than the wood itself. And if you're a purist, yes, it really isn't "solid wood" any more than plywood is, though both are made entirely of wood.

STAVE CONSTRUCTION

Door frame members can be built up from multiple staves for greater stability, then veneered.

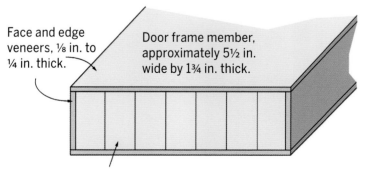

Face and edge veneers, ⅛ in. to ¼ in. thick.

Door frame member, approximately 5½ in. wide by 1¾ in. thick.

Staves, ¾ in. to ⅞ in. thick and 1¼ in. to 1½ in. wide. Grain should be oriented in same direction as veneers.

The advantages are clear for a frame-and-panel-door. If you make the rails and stiles flat and straight, they will tend to stay flat and straight no matter what wood you use. You can even make the internal staves in a more stable, or less-expensive wood, and the exterior staves in a less-stable or decorative wood. A "solid" walnut burl door is really only possible using stave construction.

At the same time, there are a few downsides to stave construction. It is much more time-consuming than milling solid wood. Oversized staves must be cut out, then glued and clamped together flat as possible. Then the glued-up stave assemblies need to be milled as if they were solid wood. Then the finished core should be veneered on all four sides. Reflect that there is a learning curve for getting stave construction right, especially if you use a mix of woods and glue line failure is always a concern.

Most stave doors use thicker veneers (⅛ in. to ¼ in.) as it makes them easier to sand the joints flush after construction and trim the door down to fit an uneven jamb without cutting through to the core material. But veneering thick wood comes with a whole range of wood movement issues that have to be addressed to avoid joint failure and delamination. Veneers thicker than ⅛ in. will behave like solid wood, with the same expansion-contraction and stability problems of full-thickness boards. If you use a thick flatsawn veneer, the way to avoid problems is to use a core of the same material, oriented in the same flatsawn direction. This way, the core will move with the veneer rather than move against each other, a fight that eventually leads to delamination.

The best stave construction uses a core of the same material as the veneer, with laminations aligning with the seams in moldings to hide their presence. And as with all veneering—make the construction symmetrical—whatever happens to one side must happen to the other (and the edges).

A compromise (not-quite-stave and not-quite solid) also works well. In this approach, you essentially create two thin frame-and-panel doors and laminate them together, the glue line interface offering rigidity that both sides benefit from. Laminating two species of wood together also works, within limitations (see Chapter 11). When I was starting out, I worked for an architectural paneling company. We paneled entire rooms, ceiling included, in solid wood. Sometimes, a customer wanted one room in mahogany and another next to it in butternut, creating a bit of a quandary for the door between the rooms. We simply laminated two ¹³⁄₁₆ in. thick solid wood doors to each other in the different woods: on one side it was a mahogany door but on the other a butternut door. This technique works well as long as the woods you're pairing have similar tangential and radial movement rates: this allows the two sides to breathe together rather than fight, warp and crack each other. Bruce Hoadley's *Understanding Wood* has a reliable chart of average rates. The biggest headache, though, is

in fitting hardware: the different mechanical properties of the woods can make it a challenge to drill and mortise across the seam.

Everything Else

By now, you're wondering "what about____?" (insert your favorite joint in the blank). I've seen (and made) doors with (in descending order of strength) shopmade floating tenons, Festool Dominos, lag bolts and rabbets, with pocket screws, dowels, even with cope-and-stick and biscuits. I've learned, however, that for full-sized doors in solid wood, none of them are lasting in the same way the pegged mortise and tenon joint is when used on its own. So I avoid these other joints except for light duty and small doors (i.e. cabinet doors) where they can work fine, or when they're integrated in a door that has other structural elements (such as plywood panels glued in their frames).

Yeah, I'm bracing for the hate mail already.

So I'll repeat the insult: each joint has a different set of strengths and weaknesses, so each has a different set of best applications. Full-sized doors are in a demanding environment and the particular qualities of the mortise-and-tenon have made it the traditional joint, and only joint (aside from its wedged variations) used on traditional doors that are still together after a hundred years.

The trouble is that we're drawn to joinery that is easier and faster to make, and hope (promise ourselves?) that it will hold together. We should think the other way—be drawn to joinery that will hold together and hope it's easy to make—even better, figure out a way to make it easily. And with patience and practice, every joint is easy to make. Yes, the mortise-and-tenon joint takes more time and effort than Dominos, dowels, or almost anything else. But not that much more time. Plus, you'll save the time of rebuilding the door in a couple of years.

And no, you don't need my permission to make doors however the heck you want. This is a free country—go for it. Maybe you'll figure out something new and create a new tradition.

Tooling

In the projects, I will show how to do the work with common cabinetmaking shop tools, sometimes adding a note on options. I've intentionally used relatively simple tools because I know more people have a plunge router than have a Virtuex lock mortiser or a Streibig panel saw. In fact, I don't have a lock mortiser or Streibig panel saw. I wish I did. But I don't: a Pity Party for me.

If a technique you find in the projects is a) done with a tool that you don't own, b) done in a way that's different from the way you've done it before, c) looks intimidating, or d) just isn't your thing, then you have decisions to make which include #1 buying that new tool, #2 doing it in a new way or #3 doing it in the way you choose, in whatever combination of familiar and unfamiliar that you like, for whatever combination of reasons you deem right, no matter how smarty-pants Mr. Purdy does it, perhaps inventing a new way along the way. If I were you, I'd always choose #3. It's wise to work within your limits to expand your limits, if that makes sense. If you're not making doors to make a living, you're making them for pleasure, so choose the approach that you enjoy the most.

A large and sturdy bench with a vise is a primary shop tool, but is actually not necessary if you have a flat and sturdy floor to work on. The basic milling of raw lumber requires a jointer, a planer and a tablesaw. They form the heart of most woodshops. Bandsaws are a key tool for cutting curves, but for any door with curved features in this book you can use a jigsaw.

To clinch nails, all you need is a hammer. I don't recommend any other tool—not your shoe, the side of a wrench or anything

Tablesaw Safety Tip

Everyone knows that power saws of every kind are dangerous as hell and you should keep all tender bits away from the blades. Everyone also knows that a riving knife or splitter on a tablesaw helps prevent kickbacks, in which the blade catches on a loose piece of wood and sends it flying back at you at 90mph. But not everyone knows that kickbacks can occur with a splitter in place, when little accumulated offcuts vibrate towards the front of the blade and pinch between the teeth and plate, sending them flying upwards past your face, one after the other, like it's batting practice time at the cage. Please don't ask how I know this. Users beware.

electric or pneumatic. You might have heard of clinching irons, and think you'll need them but you don't. Irons are used as a backer to clinch tiny nails when attaching thin strips of wood together on curved surfaces, such as in canoemaking. For a door, you bend and clinch the nails with the hammer on a bench.

Tenons can be cut using a marking gauge, a backsaw, and bench chisels, though it takes some practice and is slow going. I find the easiest way to cut them is on a tablesaw using the miter gauge and a dado blade. Tablesaw tenoning jigs hold the workpiece vertically, allowing you to cut the face of the tenon with a standard combination blade. They are, however, mostly useful for cutting smaller cabinet door frame tenons.

Chopping a mortise by hand is the age-old technique, needing only a mortise chisel and mallet. For a full-size door, though, you'll be cutting large and deep mortises and may tire of this approach. Plunge routers with mortising bits are the next step up, relatively quick and accurate. In my shop, I have a slot mortiser attachment on my jointer/planer which is great, but an expensive tool and an unnecessary investment if you're just making a few doors as an amateur.

When I started out, the shop I worked in had a huge 5HP shaper with a three-wheel power feed on it. It was great for cutting large panel profiles and molded edges quickly and cleanly. And we were all terrified of using it. When I opened my own shop 16 years ago, I did not buy a shaper, and haven't since. It's not that I continue to be afraid of the things so much as I didn't need one. I've gotten along fine with a 3HP router and router table, which are also useful for small-scale cabinetmaking. 3-D printers may be the wave of the future, but I don't own one so can't comment on their usefulness for making doors.

Specialty Doormaking Tools

If you want to specialize in doors, you will find a whole industry ready to sell you specialized tooling for making and installing them quickly and easily. Some of these tools are essential for the speed and accuracy necessary to be a cost-conscious professional. In the projects, I'll show you a few shop-made jigs that speed things along and give you better accuracy. Others simply expand the possibilities of the kinds of doors you can make. Just remember that doormaking does not require special tooling. You can make every door in this book using tools found commonly in woodworking shops.

Shapers as discussed above, are essentially industrial-sized router tables. With larger, more durable cutters, they allow you to make a much larger range of molded edges and raised panel styles.

Lock mortisers are routers integrated with a heavy duty plunge mechanism to cut the deep and wide mortises for locksets. You can also use them to cut joinery mortises.

Drilling fixtures clamp to the door and guide power drills to make accurate holes in the face and edges of doors for tubular locksets and the like.

Edge planers are small planers with a fence that tilts to 3 degrees, helpful

Shop-made mortising and support jigs make your work easier, especially when making multiples.

in cutting a door down to size and angling the lockset edge just right.

Base cutters trim the bottoms of doors at a specific height and perfectly even with the floor. They also can undercut jambs and baseboard to insert a sill plate.

Vacuum veneer presses are extraordinarily useful for making hollow core doors, as in Chapter 10, but you can always make a hollow core door like the one in Chapter 8.

Hinge templates determine the spacing of two, three or four hinges and guide a router in cutting the hinge mortises. The same jig can then be transferred to the jamb for a matching set of mortises—perhaps the most useful of all doormaking jigs. Commercial versions are generally metal, adjustable and durable, but you can also make useful jigs from scrap plywood in your shop.

Support jigs and clamps help secure a door on edge so you can work on setting hinges and hardware more easily than with the door flat on a bench. These are most useful on a jobsite where you have no bench, and are also easily shop made.

4 Hardware

If hardware is the jewelry of furniture, then doors have a problem with bling: your door can flaunt a wide selection of hinges, handles, locksets, knockers, peep holes, mail chutes, kick plates, weather-stripping, glass panes and decorative bumpers among other things. But you have to ask, as Pinkley does impersonating the General in *The Dirty Dozen*: "Very pretty, Colonel, very pretty. But can they fight?" Indeed they should be able to. Door hardware is much more than decorative accessories – it determines the door's functionality. For doors that are a pleasure to look at as well as to use, you will need to navigate many choices and balance looks with cost and durability.

Working Metals and woods

In early Colonial America, wooden hinges and locks were common. Later, forged iron became popular. But more recently, good door hardware has been made of brass. The reasons are simple: tannin-rich woods do not corrode brass; it doesn't rust like iron and steel when exposed to water and air; it is durable enough to resist long-term wear but soft enough to machine easily; it is not expensive; and it looks good—if finished it stays bright; but without a finish it gains a pleasant patina over time.

As many advantages as brass has, modern manufacturing and finishing techniques have opened up the options to include perfectly good hardware in a variety of other metals including bronze, stainless steel, zinc and aluminum. Each of these metals has a distinct set of characteristics, from weight to strength and corrosion resistance—the differences are largely unimportant in most cases unless you're after a period look, a distinctive feel or the purity of tradition—except when it comes to screws and mixing metals and woods.

Soft metal screws—the origin of many curses

If you've ever worked with brass or bronze screws, you know they are worthy of a room full of curses. Brass and bronze (and stainless steel) are relatively soft so they have to be handled with kids' gloves. Steel drive bits will burr the heads, and the screw shafts break without much warning when driving them into hard woods without an adequate pilot hole.

But every brass hinge comes with brass screws—and woodworkers who have broken a few of the heads off toss them out. Why not just use toughened steel screws that don't do this? In fact, there's a good reason that the screws are always the same metal as the hardware they secure. It's not a question of looks or cost, but of corrosion. When you put different metals in contact with each other, you create an opportunity for a galvanic reaction that eats away at the metal. Exposure to wind and rain, such as on an exterior door, exacerbates this process.

Additionally, some woods, such as white oak, are rich in tannins (perfect for wine barrels to impart that astringent taste).

Soap Your Screws

Dab the tip of brass and stainless steel screws in a bar of soap or can of wax to lubricate the threads as you drive them in. This greatly reduces the chance of snapping them. Some say this hastens corrosion of the screw, but I've yet to find evidence of this.

Rust stains streak from iron and steel hardware on high-tannin oak doors.

Tannins corrode ferrous metals such as iron and steel, producing black stains and, eventually, failed hardware. Steel screws and steel and iron hardware only mix with low-tannin woods such as cherry, pine and ash.

A practical solution is to set brass hardware first with toughened steel screws of the same size. Then, replace the steel ones with the matching brass ones. If you're working with a low-tannin wood for an interior door, then you can opt for steel hardware and not have to worry about breaking screws.

Butt Hinges

The simplicity, strength and versatility of the butt hinge makes it the king of door hanging hardware. In its most basic form, it has three parts—two leaves that swivel around a pin. Relatively small hinges can handle the shearing and tension forces put on them by large doors. And the concept has so many variations that a wide range of applications can be accommodated (Stanley publishes a useful document called General Hinge Information that runs 24 pages long). Here, we'll keep it simple and look at the most common and important options for full-size doors.

Swaged hinge: swaging allows the two leaves to close more closely, leaving a smaller gap between door and frame. This is important for weather stripping in exterior doors and

BUTT HINGE ANATOMY

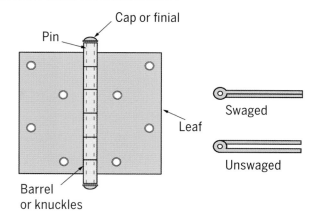

Butt hinges in a variety of configurations, clockwise from upper left: a ball bearing hinge, a projection hinge, a ⅝ radius hinge with security tab, and traditional hinges, square corners and one with ¼ in. radius corners.

sound propagation in interior doors. Most high-quality hinges for doors are swaged. Lower quality ones will not be.

Floating pins: The pins in simple butt hinges can be taken out, separating the leaves. This can be a helpful feature when hanging a door, as the process may require you to take the door off its hinges many times. However, this type of hinge is rarely of the best quality, as the fit of the pin and knuckles must be on the loose side. These butt hinges are useful for lightweight doors.

Ball bearings: Some high-quality butt hinges incorporate ball bearings on the surfaces of the knuckles. These hinges operate more smoothly and do not wear as quickly. They are consequently the better choice for heavy doors. The down side is that the pin should stay fixed in the knuckles, so hanging a door with these hinges is harder.

Projection or parliament hinges: one variant of the basic butt hinge is the projection hinge. These have leaves substantially wider than the thickness of the door. When open, the door will stand farther away from the jamb. This hinge is useful when you want the door to still be able to swing fully flat against the adjoining wall from a recessed jamb. The downside is that the hinges stick out from the jamb when the door is closed.

Square and radius corners: butt hinges often come with the option of square or radius corner. The square-cornered leaf is traditional, looks right if you like looking at your hinges, and generally appears on more expensive hinges. The radius corner generally appears on cheaper hinges. It is a convenience to make mortising the leaves into the jamb and door a bit easier if you're using a router and jig and don't want to spend time chopping out the corners with a chisel and mallet. The only difference is looks, and thereby status.

How many and how big? Figuring the size and number of the butt hinges you need for a specific door is something of a pain in the neck, as the manufacturers generally hide useful dimensions and specifications deep in the literature or online in some techno-jargon PDF.

A hinge marketed as "4½ in." usually means that both the length of the leaves and the width of both leaves when open and flat are 4 ½ in. This size hinge will generally have two leaves that are about 2 in. wide (subtracting the width of the barrel). This means they are good for a door that's 1¾ in. to 2 in. thick. From this rule of thumb, thinner doors take smaller hinges (3½ in. or 4 in. hinges) and thicker doors take larger hinges (5 in. plus). However, while many butt hinges are basically square (not including the pin or any finials),

SIZING HINGES TO DOOR THICKNESS

Size butt hinges to the thickness of the door and the width of the swing. Add more hinges for a heavy door.

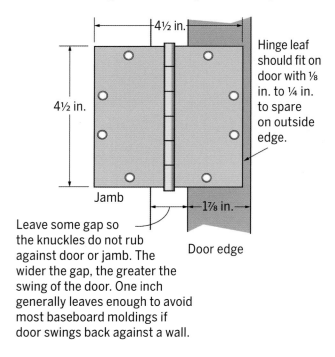

Hinge leaf should fit on door with ⅛ in. to ¼ in. to spare on outside edge.

Leave some gap so the knuckles do not rub against door or jamb. The wider the gap, the greater the swing of the door. One inch generally leaves enough to avoid most baseboard moldings if door swings back against a wall.

some are not, so always check the detailed specs to see if you're getting what you need.

It is best to mortise hinges into door edges so that the leaves are nearly to the opposite side, but not through (or you'll see them when the door is closed). This will give the maximum bearing surface between mortise edge and hinge to keep the door from sagging, but hide the hinge on the outside.

The number of butt hinges you need (or any hinge, really) for any particular door depends on the door's weight. Two is always the minimum, and three are generally best for a solid wood door. Four is better if the door is thicker, taller or heavier than average. Still, the best way to calculate the number of hinges is by comparing the weight of your door with the hinge manufacturer's recommendations.

Strap Hinges

Strap hinges mount on the outside surface of the door and jamb. They are simpler to install than butt hinges—no mortising required—but require a thicker or reinforced

Surface-mounting H-L strap hinge, spear strap hinge and larger strap hinge with lift-off pintle.

Concealed hinge that fits into the face of the jamb and edge of the door (cannot be seen when door is shut).

Double-acting hinges have a barrel on either side and allow the door to swing both ways.

jamb as they most often attach to the edge, not the face. They are also harder to adjust.

Like butt hinges, the range of quality and character varies substantially, from the cheap stamped steel ones found at your local hardware store to expensive, decorative and even hand-made ones from specialty vendors.

Strap hinges can also add to the structural integrity of a door, as the strap generally is screwed to the face of the door in several places. Still, make your doors well, so you're not depending on the strap to hold it together.

Most strap hinges have knuckles and pins just like butt hinges. But some instead have what's called a pintle, or post, on the jamb side of the hinge. This allows the door to be taken off the jamb without unscrewing any part of the hinge. And if you're wondering, there are many other types of hinges including butt hinges and overlay hinges that work similarly.

Concealed Hinges

Though more commonly used in smaller cabinet doors, large concealed hinges for full-sized doors are available and work reasonably well. Fitting them can be difficult, as they require a special jamb and precise mortising. The real question, though, is application—concealed hinges are mostly useful when you want to conceal the door.

Swinging Door Hinges

A stock slapstick gag in silent movies, the swinging door between the restaurant kitchen and the dining room always seemed to swing the wrong way, knocking the poor waiter and his tray of thirty dishes to the floor with a tremendous crash.

The advantage of the swinging door is that they never need a hand to open or close them—you can just push your way through and the sprung hinge will return the door to the closed position. A swinging door may not have many residential uses, but they're great for access between two narrow spaces

Top: Hanging door tracks mount to the ceiling for pocket doors. **Above:** Wall and floor mounted door stops prevent damage to the wall and baseboard.

Hanging tracks and pocket doors

Some doors swing, others slide. This function is particularly useful in tight spaces where a door would get in the way or hit furniture when swung open. For pocket doors, the main requirement is a wall that can be built wide enough to house it.

Door Stops

Door stops are an essential item nearly always forgotten. They keep doors from swinging open too far and letting the knob hit the wall or the door hit the baseboard. Stops come in a variety of types for different applications, and can attach to the door itself, the floor, or the wall.

Locksets, Latches, Knobs and Handles: Door Politics

By now, you should be holding your head in your hands and wondering how you'll ever make it to dinner in time if you have to sort through all the hardware options first. Perhaps you're at the limit of your despair and are now praying for a command economy in which there is only one kind of door and one kind of door hardware, period. Until that point, put the book down, get a square meal, and start up another time to sort through the nearly endless variety of hardware for keeping doors shut.

I call this section Door Politics because it's all about keeping doors closed when you want them closed, open when you want them open, and who gets through and how easily. Our politicians have the same set of issues about our borders, among other things.

Colonial thumb latches are a simple way to keep a door closed. On the downside, they tend to wear the slot in the door, enlarging it with time, and with no escutcheon, hand dirt tends to accumulate on the door around the latch. They are also are difficult to adjust so the door shuts tightly against the jamb: a door that rattles in the wind is the consequence.

where it's easiest if the door always swings away from you as you go through, such as walk-in closets or narrow hallways.

There are at least two different types of swinging door hardware. One kind is based on a pivot action, a post mortised into the floor and another in the top of the jamb. Another variety uses a double-action butt hinge.

A common safety feature in a double-swing door is a window, allowing you to see if anyone just happens to be on the other side, about to push the door towards you. I don't think residential code requires a clear glass window, but common sense suggests you install one.

Left: Outside of a Colonial thumb latch. **Center:** Inside of a Colonial thumb latch. **Right:** Traditional mortise latch has knobs or levers on either side and a simple latch bolt to keep the door closed.

Traditional mortise and surface latches have a mechanism that transfers the rotational movement of a knob or handle to move a latch bolt into the jamb, and release it. Some of these mechanisms mount on the inside surface of the door. Others are mortised into the door. The fully mortised types can be relatively difficult to install, but have great sturdiness and durability.

The standard backset of a latch, i.e. the distance between the edge of the door and the location of the knob or handle spindle, is between 2⅜ in. and 2¾ in. The wider backset gives your hand more access to a knob (your knuckles won't scrape on the door stop or jamb) but is generally unnecessary for a handle. The width of the strike-side door stile can also limit the backset you can use: thinner door stiles may not have enough room for a 2¾ in. or deeper latch set.

Tubular and cylindrical latches are the most common modern opening and closing hardware, and mostly what you'll find if you shop at a home center. They perform the same function as traditional locksets, but are designed for easy installation by drilling single large holes. Since they are so common, there is a wide variety available,

Tubular latchset looks like a traditional one on the outside, but the internal mechanism is cylindrical making it easier to install.

from extraordinarily inexpensive (and cheap) to highly durable commercial grade latches.

Locksets are simply latches with an integrated locking mechanism. Some keep the latch from moving while others drive a separate bolt. Surface bolts and deadbolts are locking mechanisms that are separate from the latches. A locking function adds expense to the hardware, but can be useful to keep people out of rooms they shouldn't be in, or perhaps in rooms they want to leave; but this touches on the political side of doors so I'll leave the issue up to you.

Knobs and handles are the two basic choices of how you work the latch mechanism.

The choice is largely an aesthetic one, with knobs generally looking more traditional; but the social trend towards greater accessibility is giving the handle the upper hand. Put simply, knobs require good hand strength to open whereas handles need only weight put on them. It is far easier to open a handle than a knob if you are disabled, carrying two bags of groceries, or a child.

One important installation note—in a frame-and-panel door, latches and locks should not be aligned with a rail, as they require a large mortise that will weaken the joint. The heights of latches, locks and handles are determined by the building code (see earlier), so design your door with hardware offset from the rails as much as you can.

Knockers and Bells

Without a knocker or a bell, your visitors will be tempted to pound on the door to let you know they'd like to come in. For some reason, you will rarely find a traditional knocker or bell located anywhere other than the vertical centerline of the door. And so that's where people look to find them. Locate them elsewhere at your peril. Door knockers seem to be one of the few pieces of door hardware that are available in not just understated elegance, but also expressive creativity.

Weatherstripping

Weatherstripping creates an essential seal between your exterior door and jamb that keeps cold air out in the winter and warm air in, in the summer. It also keeps out water, bugs and sound. It's this last feature that makes weatherstripping also useful for interior doors.

Vinyl weatherstripping that you nail to the jamb is the most common. You'll find it at every home center. It works fine, though is not particularly durable and will need replacing when it cracks. It also may look cheap on a handmade door.

Left: Integrated locksets have a latch bolt and deadbolt that can be keyed for locking the door.

Below: Door knockers range from the understated to expressive and creative.

Wood door stop molding can be kerfed on the inside corner to fit a vinyl or silicone rubber tube style weatherstripping. This is the most hidden style and is very effective.

Metal weather stripping is far more durable. The most common are interlocking bronze strips, often v-shaped in profile. The downside is the fussiness necessary to install it right and the imperfect (albeit good) seal it creates. Metal weatherstripping can also rattle in wind, unlike the flexible varieties.

The underside of a door requires a different type of weatherstripping called a sweep. The sweep has to be durable and flexible since it does just that—sweeping the saddle every time it opens and closes. Some sweeps use windshield-wiper like strips of rubber silicone, others use fine bristle brushes. Others are automatic, descending down only just as the door closes.

Glass for Doors

Glass in a door must be tempered or laminated. It's a safety issue to prevent injury if the glass ever breaks. Remember—doors get slammed no matter how often you remind the kids.

Laminated glass will crack, star and shatter, but it won't come apart: making a hole in it takes a fair amount of effort, a benefit if you'd like to slow down anyone trying to break in. This is the type of glass they use in cars windshields.

Tempered glass shatters into thousands of relatively harmless pieces, not into long sharp shards the way plate glass does. It breaks relatively easily and completely: strike it with a fine point on one corner and the entire sheet will explode. This is what they use in car door windows, making it really easy to get a person out in an accident.

These types of glass are often sold as "safety glass," and there are many different types and grades. Some are bullet proof.

Insulated glass, in which two panes are sandwiched with an inert gas filling the space between them, can increase the R-value of an exterior door. The amount of glass that you have in the door will determine how much R-value you gain by using insulated glass, but generally it is not much.

Ordering glass for a door either from a local shop or online is pretty simple. The quality is pretty much the same as they buy glass from a few central manufacturers. But glass shops do differ in terms of service and the accuracy of the dimensions they cut. My advice is to make two matching templates for the glass you need, sizing them a wee bit smaller than the groove or rabbet they will sit in. Keep one template, so if there is any question about a wrong shape or size, you can produce it. Templates given to glass manufacturers universally disappear in the process.

Other Hardware

As there is probably no end to the hardware options for doors, it's best to end this section here to avoid a book heavier than a dictionary. Let it just be said that, if you need hardware for some function, it probably already exists, or you can create it on your own. Dog and cat doors abound, though integrating them into a wood door takes some special engineering to preserve the door's durability and its R value. Peepholes for scrying the person on the other side are a fun touch for both interior and exterior doors. Mail chutes are a necessity in some cities, though in the country where I live the postal service never gets closer than my mailbox on the street. Kick plates will make a high traffic door look protected and reduce wear. They sometimes trap moisture, though, helping speed rot in a lower stile, so it's key to apply the finish to the door before the kick plate.

Finger Guards

If you've read this far, you must be really into doors, have nothing else to do, or are looking for an Easter egg, like those in movies that put some after the credits. OK, here's something for you: finger guards. If you think America is hyper risk-averse, this is fodder for jokes. But if as a kid you pinched or lost a finger in a doorjamb (it happens more often than you might think), and you don't want the same to happen to your kid, then installing finger guards all around your house might just be for you. These are flexible strips of plastic that go over the joint between door and jamb. When the door opens they expand making it impossible to fit your fingers, large or small, between the door and the jamb before it closes.

5 Hanging Doors

A well-hung door is a matter of good planning and attention to detail. It's best learned by working with someone who has a few years of experience, as the difficulties are in troubleshooting, and troubleshooting can't really be put into a linear description or photo essay. Nevertheless, below is my basic process, taking you through the hanging of a simple door using butt hinges. You'll find that you have many options: the key is to consider them before you begin work. Yeah, yeah, I know: it's the same good advice that we all ignore, like reading car stereo instructions before installation. Yes, you can figure a lot out as you go along: but when you have a problem, I get to say told you so.

Milling the Parts

For a typical interior jamb for a rough opening (i.e. new construction) you'll need two side jambs, a head jamb, casing for both sides, side and head doorstop and possibly a sill. If you're making a door for an existing jamb, well, then you can skip this part.

For most lightweight interior doors, the jamb can be made from 4/4 material, milled $^{13}/_{16}$ in. thick. It can be the same wood as the door or the wall (painted if the wall is sheet rock). The jamb should be as wide as the distance between the finished walls. The casement's job is to cover the gap between the jamb and the walls, generally an inch or two. The doorstop's job is, well, obviously—preventing the door from closing too far and springing the hinges. It also creates an air seal and provides purchase for weatherstripping.

The jamb and door stop are usually flat, square stock, as most ornaments would interfere with the function. The casing, however, is usually molded, sometimes ornately. One consideration is the thickness of the casing. If it is too thick, it can prevent the door from opening fully.

Basic Door Jamb and Casing Parts

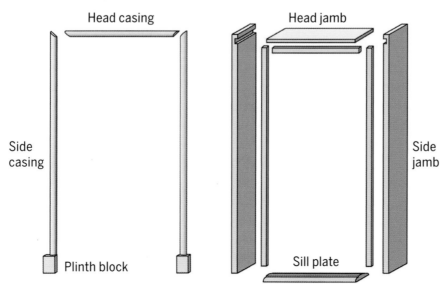

Head casing

Side casing

Plinth block

Head jamb

Side jamb

Sill plate

The plinth block is a handy feature to enable a simple transition from casement to baseboard. It should be larger than the casement and baseboard with two flat sides for each to terminate in it.

Start with Measuring and Sketching

Making a full-size plan view drawing of your door and jamb is an extraordinarily useful first step. It will answer many dimensional questions, serve as a touchstone through the building process and reduce errors.

Start by measuring the door opening between walls and from floor to ceiling. Importantly, measure the diagonals to see how far out of square the opening is. You don't have to make an angled door to fit an angled doorway (unless you're recreating a set for the Batman TV show from the 60's), but you will have to make a smaller rectangle to fit in an angled opening.

A section through the door and jamb in plan view is the most useful drawing. If you don't have a sheet of paper roughly 1 ft. by 4 ft. (and I don't), you can tape a few pieces together or, use a cut off from a sheet good or a cardboard backer. Then again, if you're part of the digital generation, you might be more comfortable making this drawing in a CAD program such as Sketchup.

RO Sanders Safety Tip

Random orbit sanders are among the best power tools. They save a lot of hand sanding, leave a great finish and are relatively safe. I figure that they don't even appear in emergency room statistics. However, they do depend on a little plastic ring to prevent the pad from spinning at supersonic speed when the sander isn't on the workpiece. If that ring wears out, setting the sander down on your workpiece when it's running will cut a deep gouge into the surface. And if you set the sander down on the workbench just a little too close to a rasp, it can send that rasp clear across the shop and break a window. Please don't ask how I know this. Users beware.

Which Way to Swing

Commercial exterior doors must swing out. It's an emergency consideration: if a hundred people are trying to force their way out of a building, they could jam up against an in-swinging door and prevent their own escape. However, there is nothing in the code about swing for residential doors. So it's your choice to swing in or swing out.

Some guidelines: most residential exterior doors swing in to allow for an outward-swinging screen door. For interior doors, the best swing direction is sometimes obvious and sometimes not. Bedroom and bathroom

JAMB AND CASING IN SECTION

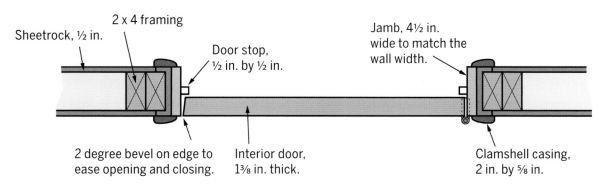

Sheetrock, ½ in.

2 x 4 framing

Door stop, ½ in. by ½ in.

Jamb, 4½ in. wide to match the wall width.

2 degree bevel on edge to ease opening and closing.

Interior door, 1⅜ in. thick.

Clamshell casing, 2 in. by ⅝ in.

doors should open in, so as not to block the hallway when left ajar. When open, they should swing back against a wall, not in front of another doorway. Closet doors generally open out because the closets are too small to allow the door to open in (though large walk-in closets are an exception). Doors between rooms are often a toss up. Looking to see what the open door would block can help the decision. And yes, prying questions about "which way do you swing?" is a common form of doormaker harassment on jobsites.

Door latches sometimes have a 'handedness', which is to say they are set up differently for the side the door is hinged on and the way it swings. To get the right hardware, it takes a moment to figure out what 'right hand in-swing' is as opposed to 'left-hand out-swing.' The key is that your point of view should always be "outside" to keep the left and right straight.

Make the Jamb and the Door Together

Some make the jamb first, then a door to fit it. Others make the door first, then a jamb to fit it. I recommend designing and making both together to avoid the compromises necessary when one has to fit the other.

Once completed, attach a temporary sill, or the real one, to the bottom of your jamb uprights and hang the door in your shop. If the saddle has a profile, often a gentle rise, cut the bottom edges of the jamb to conform to the shape. If I'm making a door for my shop or my house, then I'll skip this step, since I don't have far to go to get tools. But if the door is to be installed at a distance, pre-hanging is an extraordinarily useful step to avoid repeated trips back to your shop.

Sills—wood versus metal

Most interior doors do not need saddles—in fact they're trip hazards so it's best to make sure the floors are continuous between rooms (a flooring book issue). Saddles can be necessary between rooms that have different floor coverings, such as from the carpeting of a bedroom to the tile of a bathroom. These saddles should be flush with the floor; but if for any reason they can't be, as carpeting must fit under them, then make the rise as minimal as possible.

Exterior door saddles (or thresholds or sill plates—they seem to have a lot of names) are another matter entirely. They are often raised as their job is to keep wind and water out, helping create a seal with the bottom edge of the door when shut.

I'm often asked for a "durable wood saddle," the implication being one that will stay looking brand new like the door forever. Yes, oak is durable, and yes, I can apply a durable finish on it. A flooring finish will be extraordinarily hard, but they are designed for interior use and will not stand up to the wild temperature variations the flexible exterior finishes are designed for. These flexible finishes, however, are soft by nature, so wear away quickly when constantly stepped on.

This is why I recommend commercial metal saddles for exterior doors unless you like the worn look or constant maintenance.

Hinge Mortising

There are two basic ways to cut the hinge mortises. One is by hand, the other is using jigs. The advantage of the jig is speed and accuracy. The disadvantage is that you have to make or buy the jig.

Refer back to the illustration of sizing butt hinges in Chapter 4 to understand the basic layout of how butt hinges should fit a door and jamb.

In summary, a swaged hinge should be set flush with the face of the door and the face of the jamb, so the depth of the mortise is the thickness of the hinge. The length of the mortise is the same as the leaf, and the depth is set by the thickness of the door.

Scribe the hinge sides on the door edge, knuckles tight to the door face.

Use a marking gauge to scribe the width of the hinge on both door edge and jamb.

Use a chisel to mark the mortises edges on the face of the door with a deep cut.

Rout the mortises with a straight bit, keeping within the layout lines.

Use a wide chisel to chop and pare the mortise corners square.

By hand, start by propping the door up on edge, either clamped to your bench or using a support jig. Drape the hinges on the door edge where they will sit, 5 in. to 7 in. from the top edge and 9 in. to 11 in. from the bottom edge and scribe the location of the top and bottom edges.

Set a marking gauge to the width of the mortise (door thickness minus 1/8 in. to 1/4 in.) and scribe between your layout lines (leave this setting on the marking gauge and use it when you lay out the hinges on the jamb faces).

Mark mortise ends on the door face with a knife or chisel, just the thickness

of the hinge. Make a deep, accurate cut as this acts as a chip breaker for the router.

Chuck a straight bit in a small router or laminate trimmer and set the depth to the thickness of the hinge. Rout the hinge mortises by hand, staying within your layout lines.

If you're using radius corner hinges, the bit should match. If you're setting square hinges, chop the corners of the mortises square with a chisel.

Lay the hinge jamb over the door edge, aligned so the top of the door is just below the rabbet, and transfer the mortise locations.

A long level helps ensure the jamb is plumb and straight when you attach the hinge side first.

Attach the jamb to the wall using long screws set in the hinge mortises to hide them.

There are jigs that do guide the router to make each mortise precisely the same. Buy or make one if you plan on making many doors.

Set the hinges in the jamb and door mortises using only two of the four screw holes. This allows you to adjust the hinge's location when you hang the door. Remove the hinges from the door and jamb.

Setting the Jamb

Set the assembled jamb within the rough opening, aligning the hinge side with the interior wall, or where the interior wall will be.

Check and shim the bottom up if necessary to reach the correct height. Set the jamb side in place with screws through the jamb under the hinges or where the door stop will be located. I use screws because they are adjustable and reversible, unlike nails or adhesive. As needed, shim the back side of the jamb to make it plumb and straight, also 90 degrees to the interior wall.

Align the strike face side of the jamb with the interior wall and attach it to the rough opening. Shim it to the proper height so the jamb is a rectangle. Be sure to avoid twisting the jamb face and keep the top of the jamb level.

Check the level of the head jamb before attaching the latch side of the jamb to the wall.

Ensure the latch side of the jamb is plumb and flush with the wall before attaching.

Check the gap between the door and jamb to ensure it's even all the way around. Shim the jamb as necessary.

Hang the Door

Shim the door to the right height in the open position, align it with the jamb and attach the hinges to hang the door. If the hinges have a removable pin, this process is much easier.

If you've pre-hung the door, it should swing shut quite easily. It's now a matter of adjusting the screws that hold the jamb to make the gap even all the way around.

If the door is not pre-hung, then you may need to resize the door to fit the jamb. Take it off the hinges to do this work. To run a belt sander or planer along the edges of a hung door is both dangerous and stresses the hinges.

In either case, the goal is an even gap, around ⅛ in., around the three sides of the door and a little more on the bottom edge.

A Slight Taper on the Strike Edge

Most doors work best if the strike edge has a slight taper along the front. This allows them to clear the jamb without rubbing and still have a small gap when closed.

When you first swing the door closed, check this gap and adjust with a block plane where it rubs or gets too close. A block plane works best for this, as the gap is rarely the same along the entire length.

Fit the Latch, Other Hardware As Desired

Some latch hardware comes with templates, even instructions, to locate it correctly, but most do not.

Traditional latches should be centered in the door so that the mortise walls on either side are the same thickness. Once you've mortised the latch body into the door, swing it against the jamb to find the right location for the strike on the jamb. Be sure to measure the proper distance in from the edge of jamb, so that when closed, the door will align with the jamb edge. Cut a mortise beneath the strike plate that is larger than necessary. This allows you to adjust the strike location easily.

Troubleshooting

If your door closes and opens easily on the first try, then bravo! Go relax with friends. But if not, then you are a mere mortal like the rest of us and need to start the troubleshooting process. Here are a few adjustment tools:

I find it's always best to sit down and think the problem through before you start tearing things apart and changing them. You might know exactly what the problem is, then again you could be wrong. Frankly, this is the one good use of having a smoke on the job: it forces you to stop for a moment and offers the insight of reflection. But the demon cigarette is no longer fashionable, so have a bottle of kombucha and a bran cracker instead.

Perhaps the solution is to plane the face of the door down in one spot—are you glad you haven't applied a finish to the door yet? You bet.

Perhaps you need to move the hinges up, down, in or out. This is why you hung the door using only two of the screw holes.

Reset the hinges in two new holes and plug the old ones with the ends of shims or wound up cardboard. The bits of shim can also be glued behind the hinge.

Also, don't forget the screws holding the jamb in place. They can pull the jamb in or push it out as needed. You should start with plumb and square, but feel free to move away from it to make the door work better. But be careful not to go so far out of plumb that the door swings itself shut or open.

Set the Door Stop

With the door closed, attach door stop trim from the outside (hopefully over the screw holes that are holding the jamb in place). If you're not using weatherstripping, I place the trim gently against the door, using a business card to get the right gap, and either nail or screw it in place. Be careful that the molding does not squirm either tighter against the door or away from it. In the upper strike side corner, I apply the trim without a gap.

A door should be easy to close without pushing, yet not rattle when closed. To do this, the door should touch the top of the stop first, before the latch goes home in the strike. This little bit of tension holds the door in place without rattling.

Finish the Door and Jamb

Once the door works just right, take it off its hinges, take off its hardware, steam out any dents, sand the rough edges, corners or surfaces, apply the finish of your choice and rehang the door. You're done! Right? Well, no—look at the gap between the wall and the jamb….

Casing

The practical purpose of casing is to cover the gap between the jamb and surrounding walls in a decorative manner. Casing can be as simple as three pieces of clamshell molding or as complex as a classical entablature with architrave, frieze and a cornice. How you should case your door

Larger casings can be built up from smaller moldings.

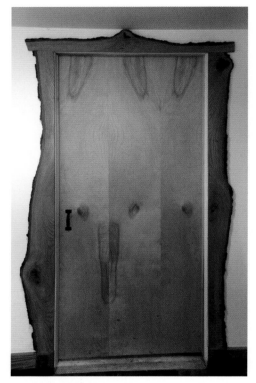

Natural edge board offcuts make door casing with character.

frame depends on the style of the door and the house its going into; also your patience.

While many lumberyards sell door casing in a wide range of woods and profiles, making it on your own is a good option. You can make it from the same boards as the door and you're not limited to commercial profiles or even the size of your router bits. Larger moldings can be built up from smaller profiles. If you can get a hold of natural edge boards, the offcuts make a rustic casing that

Butt the vertical casement into the top one for an Arts-and-Crafts look.

can dress up otherwise plain doors. While most casings are mitered at the top corners, they're not in the Arts-and-Crafts style.

An often overlooked key to interior casing that looks right is to remember that the door casing should integrate well with the baseboard. As baseboards and door casings have different profiles, it is not a given that the two will match easily. Baseboards tend to be wider and may have a base cap and or a shoe molding. Ideally, these end cleanly against a flat-edged door casing that runs to the floor. However, this requires the door casing to be thicker than the baseboard and flat on the edge. As profiles with these features will not always look good running around an entire doorway, a good solution is a separate plinth block. This allows you to choose the door casing with no concern about how it integrates with the baseboard.

And if all this isn't enough to get you where you need to go, a more thorough reference to hanging doors is Gary Katz's *The Doorhanger's Handbook*, Taunton, 1998.

PLINTH BLOCKS

Plinth blocks solve the problem of the intersection between the baseboard and the door casing. They need to be wider and thicker than the casing and taller and thicker than the baseboard.

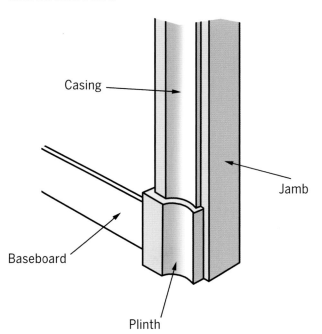

Casing

Baseboard

Jamb

Plinth

Belt Sanders Safety Tip

When I was working in an architectural millwork shop in the 80's, we still used belt sanders—big, honking 4 x 24 belt sanders heavier than anvils. Touching the running belt rarely hurt (though goodbye tan in that spot), but the opening where the belt goes back up into the housing was a dangerous spot. With 80 grit, those sanders were strong enough to catch and tear off a loose shirtsleeve end. And they were also powerful enough to catch and jam loose fitting pants and a hank of skin above the knee pretty good, that is if you let the sander swing down off the bench after a pass. Please don't ask how I know this. Users beware.

Door Projects

In this section, I'll show you how to make a variety of basic residential doors, covering fundamental construction techniques in traditional and modern designs. Of course I'll make them in some of my favorite styles, but I'll keep it simple so you're not distracted by a lot of details or features. You'll find the wild, the masterful and the inspirational in the gallery. Nevertheless, it's my hope that there will be nothing in the gallery that you can't make using the basic techniques found in the project chapters (well, maybe a few things).

Each project shows the making of a "slab." This is industry terminology for the door, only the door and nothing but the door. I address which wood, hardware and finish I used for each door, and address any special hanging concerns, but I don't go through the step-by-steps for all of it. These were discussed briefly in earlier chapters, and the best thorough knowledge of each can be learned through more focused books.

As most frame-and-panel doors have similar construction, I've avoided repetition by referring to processes discussed elsewhere. **Chapter 2** has a detailed how-to on milling solid lumber straight and true. **Chapter 7** serves as the most detailed example of mortise-and-tenon joinery. Some chapters will offer complete how-to directions for making

The lock fit, but then the bolection molding didn't: mistakes are design opportunities.

the door slab, while others will focus on a detail or two, referring to the other chapters. Again, this is in part to avoid repetition, but also because some of the doors were already made when I began the book. I don't recommend you use this book for Instant Door Recipes. Doormaking (indeed all things requiring craftsmanship) doesn't work that way. If you want to build one of the doors,

ANATOMY OF A DOOR
Doors have a wide range of parts with particular terminologies.

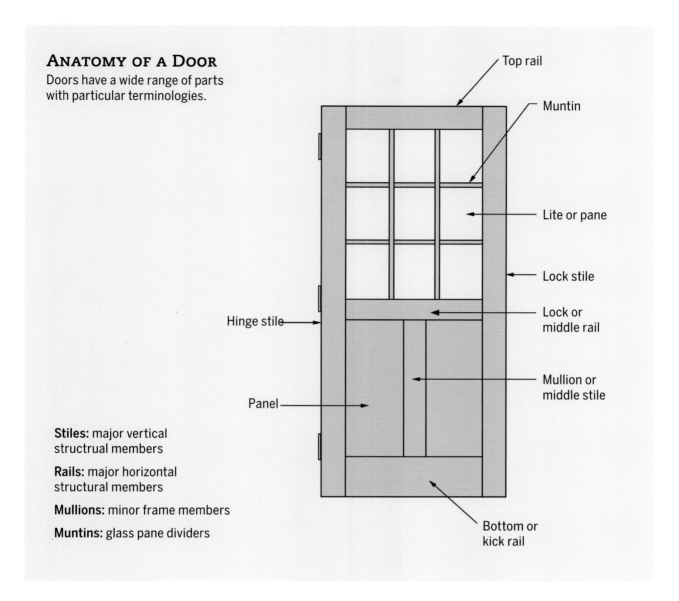

Top rail

Muntin

Lite or pane

Lock stile

Lock or middle rail

Mullion or middle stile

Bottom or kick rail

Hinge stile

Panel

Stiles: major vertical structrual members

Rails: major horizontal structural members

Mullions: minor frame members

Muntins: glass pane dividers

please read and absorb the whole book first. This process will make individual instructions (especially where there is no image) so much more understandable and allow you to adapt the instructions to your specific circumstances.

Indeed—I hope and trust that you will adapt these door projects to your own tastes and circumstances. If the door calls for 6½ in. wide black locust stiles but your lumber yard only has 6 in. wide African mahogany, I promise your door will come out just fine. If you're unsure how thin a stile can go before it's too thin, well, I'm not sure either. One inch wide is definitely too thin, but I've made lasting doors that were 3 in. thick.

Board-and-Batten Door

6

Board-and-batten doors are a mainstay of the oldest and simplest Colonial American homes. The originals were most likely painted and used a bit of rope or a thumb latch to keep them closed. They are rustic, light-duty and simple to make. On the other hand, they're about three steps up from animal skins stretched over a stick frame. Putting them in your house may solicit accusations of advanced camping as they're thin, don't insulate well for sound and heat and are not too dimensionally stable. But they look nice. So they are great where insulation and stability are not primary considerations, such a pantry door, closets or the shed out back.

There is a definite front (the boards) and a definite back (the batten). You can butt the boards together, but over time they will shrink and leave gaps. It's better to ship lap or tongue-and-groove them. The traditional way to secure the boards to the battens was with clinched nails. These can be decorative if you use roseheads or cut nails. The more modern alternative is to use screws which can't be seen (or shouldn't) from the front.

Board-and-batten construction is not particularly rigid, allowing the members to twist relatively easily. Exterior doors won't seal well against the jamb. You can clinch nails through most hardwoods, but pine is easier to work with and more traditional. A board-and-batten door in walnut is a bit like making a pair of jeans out of silk.

Variations

A good variation on this door is to use thicker wood, either 5/4 or 6/4, in a species with a little more interest such as butternut or chestnut. A door using 5/4 boards and battens could still be made with 3 in. nails, but I'd use longer ones, 3½ in. or 4 in. for the 6/4 door. Using heavier stock will also increase the door's rigidity. You might also want to add a third horizontal batten and add two diagonal battens between them.

Using wider and fewer boards is fine as long as you make deeper grooves and wider tongues. The movement across a 12 in. board can be substantial, so expect gaps as large as ½ in. in the driest part of the year unless you'd like to peep through your door. I wouldn't use fewer than three boards. Using more boards can look good as well, just adds a lot of extra work.

Winnowing the principles of this door down to the basics, realize that you're just connecting using metal fasteners to hold boards together in a way they can still move. For larger doors or using heavy woods, you can use screws or carriage bolts.

BOARD-AND-BATTEN DOOR

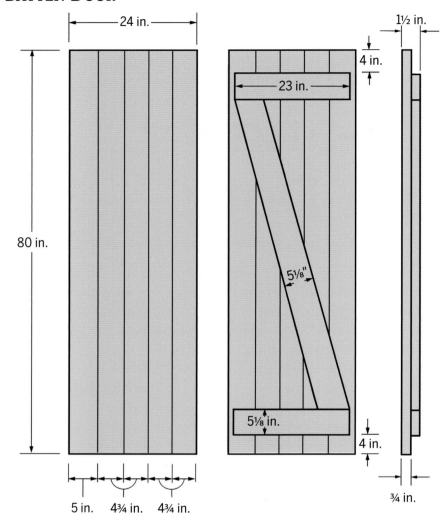

24 in.

80 in.

5 in.   4¾ in.   4¾ in.

23 in.

4 in.

5⅛"

5⅛ in.

4 in.

1½ in.

¾ in.

FINISHED DIMENSIONS:
24 in. by 80 in. by 1½ in. thick

MATERIALS
- 6 @ 1 x 6 by 10 ft. common pine
- 1 lb. of 3 in. rosehead nails (about 35–40 count)

CUTLIST
- Boards 5 @ 81 in. by 5⅛ in. by ¾ in.
- Battens 2 @ 23 in. by 5⅛ in. by ¾ in.
- Batten 1 @ 70 in. by 5⅛ in. by ¾ in.

HARDWARE
- Strap hinges
- Colonial latch

Step 1

Pick out six relatively flat 1 x 6 by 10 ft. pine boards at the lumberyard. I use the common grade because this is a rustic project and I'm not painting it. For a painted finish, I'd use select boards to avoid large knots, as the sap can seep through paint, no matter what Zinsser claims.

Step 2

Cross cut five of the boards 81–82 in. long, just over your finished dimension. You'll trim the door to final length after construction.

Step 3

Rip the boards 5⅛ in. wide on the tablesaw.

Step 4

Lay out five boards in an arrangement you like. I put the straightest boards with fewest knots on the sides. Mark them so you know which edges get a tongue, which edges get a groove and which edges get nothing. Number the boards or draw a triangle across them so you can put them back together in the same arrangement later.

Step 5

Use a ¼ in. wide slot bit to cut the grooves on all the boards, but only on the edges marked "groove." If you're wondering why I stress the obvious here, ask me again after you cut a groove on an edge with no marking. The groove should be between ⁵⁄₁₆ in. and ⅜ in. deep, but no more. The exact width of the groove is not important, but it should be centered. I get the bit approximately centered, cut the groove from the face of the board first, then flip the board over and cut the groove again. This approach produces a groove about ⁵⁄₁₆ in. wide and perfectly centered.

Step 2: Use a circular saw to cut the boards to rough length. Clamp the boards to the bench so they don't move.

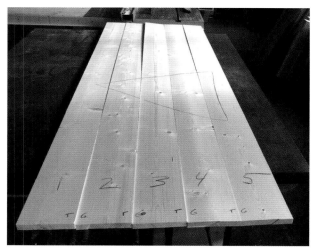

Step 4: Mark the boards to indicate their order and which ones get tongues and grooves.

Step 5: Cut the grooves with a router and ¼ in. slot bit, centered on the board.

Step 6: Cut rabbets on both sides of the board to create the tongues.

Step 7: Chamfer the tongue edges so they can slide easily into the grooves. At the same time, chamfer the face edges to hide the joint.

Step 9: Use a framing square to lay out the locations of the end battens, about 4 in. in from the bottom and top edges.

Step 6

Use a rabbeting bit to cut ¼ in. wide rabbets on both sides of the edges that need a tongue (sometimes you can use the same slot bit to cut your rabbets, but I favor a slightly deeper groove than tongue). First, set the depth of cut a bit shallow and test the fit. You want the tongue and groove to fit snugly. Then make slight adjustments to the cutting depth to fine tune it. Remember that each adjustment is doubled as you cut from both sides. You can cut both the tongues and the grooves on a tablesaw, but it's not nearly as accurate if the boards are warped. When you re-read some of this technical how-to language outside of a woodworking context, especially about rabbets and tongues, it sounds really funny. Try it on your friends.

Step 7

Lightly chamfer the edges of the tongues so they go in the grooves easily. At the same time, chamfer the front side edges of the boards if you like. I prefer a light ¹⁄₁₆ in. wide chamfer between the boards to hide the joint. More often you'll see a heavy, nearly ¼ in. chamfer. This is a purely aesthetic feature to hide the joint, so chamfer as you like.

Step 8

Clamp the boards together to test the fit. At this point, I sand the fronts of the boards to 220 grit before assembly, as you can't sand them with the rosehead nails in place. In this particular door, I didn't sand the interior surfaces since they'll be inside a closet; but by all means sand the backs if you prefer.

Step 9

Lay out the locations of the two end battens (also sand them, if you like) on the back side of the door. I put them about 4 in. from the top and bottom.

Step 10: Drill pilot holes through both batten and boards, spaced evenly.

Step 10

Clamp the 23 in. long battens to the door and drill pilot holes for the clinched nails all the way through the battens and boards. Where did the battens come from? The offcuts from the boards, of course. That's why you start with 10 ft. long ones. Locate the pilot holes at least 1½ in. from any edges to reduce the chance of splitting.

Step 11

Hammer the nails in partway from the front so that about ½ in. sticks through past the batten. The 3 in. nails should stand about ¾ in. proud, and don't hammer them any deeper. As you don't want to nail the door to your bench, or in my case bend the nails against my steel bench, use the other offcuts as backers. No worries—if you accidentally nail the door to the backers, they are easy to pry off.

Step 12

Flip the door over and bend the nail points flush with the battens, hammering them from a strong angle. The door will stand on the proud nail heads, which is fine. Nails bend easily, and this is one situation in which bending a nail over is a good thing.

Step 11: Hammer in the 3 in. nails from the front, leaving them proud by about ¾ in.

Step 12: Bend the nail tips flush with the battens by hammering from a strong angle.

Step 13: Hammer the nails all the way in from the front, a little proud of the surface.

Step 14: Clinch the nails flush with the batten by hammering from a strong angle.

Step 15: Use a bevel gauge to find the angle for trimming the end of the diagonal batten.

Step 13

Flip the door over again (by now you've noticed you do a lot of that when clinching nails) and hammer the nail heads all the way in. I like to leave the roseheads a little proud to give the front some texture, but hammer them flush if you like.

Step 14

Flip the door onto its face and now hammer the bent points of the nails sideways again, clinching them into the battens. This is the trick that will keep them from ever coming back out again. Do note that there's nothing pretty about clinching a nail. You'll get tearout, the occasional split and hammer marks (if you're not perfectly careful), all of which are part of the rustic appeal, and not mistakes the way they would be in a more refined door. So go rough on the door and easy on yourself.

Step 15

The third batten crosses the boards diagonally, giving the door some resistance to twisting that a third horizontal batten would not. The top of the diagonal batten should be on the lock side and the bottom on the hinge side. Oriented this way the batten also helps prevent sagging, i.e. the door turning into a parallelogram. If you have greater than usual twisting or sagging concerns, you can add a second diagonal across the first, making an X pattern; but this more work and weight on the hinges. Use a T-bevel to find the correct angle to trim the ends of the diagonal batten. I use the miter gauge on my tablesaw for this angled crosscut, but chop saws, handsaws, and circular saws will also work fine.

Step 16

Attach the diagonal batten to the boards with clinched nails, in the same way you did the two horizontal battens. Locate the nails at least an inch in from the edges of the battens and the boards. I find it easier to lay out the location of the batten on the door front to ensure I'm not missing the boards or battens, and drill from that side. Lots of flipping, yes, I know.

Step 17

Crosscut the finished door top and bottom to length with a track saw. I find these tools extraordinarily useful for jobs like this, in spite of their expense, as they give a chip-free and accurate crosscut.

Step 18

Prep the door for finishing with 180 grit sandpaper. A soft eraser works well to get rid of pencil lines before sanding. I used a single coat of Danish oil, following the directions on the can. The oil soaks into the wood for a matte finish that does not protect the door from wear much at all, allowing it to gain a natural patina from use pretty quickly—perfect for a rustic door. The oil also darkens the nearly white pine to a gold that will continue to darken with time.

Congratulations—you're now the proud parent of a door slab.

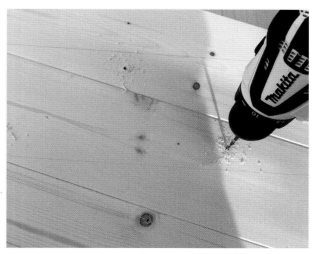

Step 16: Drill pilot holes from the front for the nails that clinch the diagonal batten.

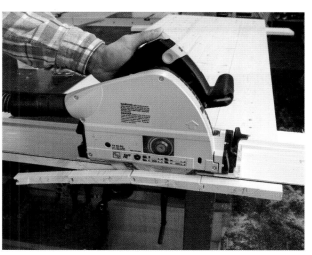

Step 17: Cut the entire door to length with a plunge saw and track.

Frame-and-Panel Interior Door

7

While there are many good, and great, ways to make a door, this is my favorite—not to bias you against it if you're contrarian by nature. Making these was how I learned doors thirty years ago in an architectural millwork shop, Wood Interiors by Rodger Reid. And I still make them in pretty much the same way.

Solid wood frame. Pegged mortise-and-tenon joinery. Molded frame members (moldings not applied later). Mitered molding corners. Floating solid wood panels. If you make this door reasonably well—and this design tolerates a lot of mistakes and almost-rights—it will never come apart unless intentionally taken apart. It is dimensionally stable within 1/64 in. in most woods. And it looks good. If there's one door you need to learn how to make, this is it.

The techniques you'll find in this project are foundational, in that many of the other doors are built with the same techniques. I'll refer back to this chapter for details in later projects. You'll need the major shop tools to make it: jointer, planer, tablesaw, plunge router, a solid bench and basic hand tools.

This door is made from leftover pieces of beech from another job. The kiln had dried the wood too fast leaving huge internal checks. But where there's a problem, there's an elegant solution—clear epoxy

fills that both restore the strength of the wood and add a beautiful feature—and yes, light shines through the epoxy.

Of course, use whatever wood you like or can find—I can't guarantee a source of badly dried beech with a lot of character, but common grades of lumber will often have defects that the epoxy fill technique improves.

Variations

Look around in North America and you'll find the frame-and-panel design everywhere. Even aluminum garage doors are stamped with the design. Winnowing the principles of this door down to the basics, it's a rigid frame that does not change much dimensionally (realize the distance between the pegs does not change) with captured panels within the frame that do change dimensionally. The panel configurations and moldings are all that separate Colonial from Victorian, Arts-and-Crafts, Shaker, Modern and beyond.

FRAME-AND-PANEL INTERIOR DOOR

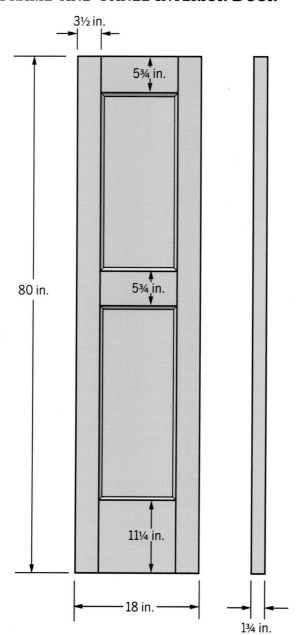

3½ in.

5¾ in.

80 in.

5¾ in.

11¼ in.

18 in.

1¾ in.

Milling and Shaping

This section offers basic information for creating the basic parts of a frame-and-panel door. I'll refer back to it in subsequent chapters, discussing the differences as necessary.

Step 1

Mill your rails and stiles following the steps in **Chapter 2** for **Rough Milling and Fine-Tune Milling** to end up with two stiles, straight, flat, square that are 3½ in. wide by 1¾ in. thick and at least 80 in. long. At the same time mill the rails to the same thickness, all a little long if possible. As they are shorter, they may not need to be fine-tune milled the same number of times, but be sure they end up the same thickness as the stiles and just as straight and square.

Step 2

Clamp the stiles together and lay out the mortises on both at the same time. Mark where the rails intersect the stiles. Instead of measuring I often lay the rails themselves over the stiles: make sure they're square and scribe the lines.

Step 3

Mark the length of the mortises in the stiles, clamping them together to ensure the layout marks are the same on both sides. The mortises should not be the full width of the rails, but ⅝ in. in from either side to allow for the panel grooves. At the ends of the stiles, they should be at least 1 in. from the top and bottom of the stiles. The large bottom rail should be divided into two tenons, ½ in. between them.

Step 4

Mark the width of the mortises with the marking gauge. Scribe from both sides to ensure you've centered the mortise. Make the mortise a little more than half the thickness of the stile, about ¾ in. or ¹³⁄₁₆ in.

Step 5

Cut the mortises in the stiles about 2½ in. to 2¾ in. deep. Really, the deeper the better. Of course, the depth depends on the tools you have or are willing to buy. It doesn't matter which tools you use to cut the mortises (hammer and chisel works fine, though it's a lot of work); but the depth does effect the strength of the joint. The strongest joint would have the mortise ⅛ in. less than the full width of the stile, but I don't have a router bit that long, and can't find one.

Step 2: Lay out the locations of the rails on the stiles. Use the rail itself to get an accurate width.

Step 3: Mark the ends of the mortises ⅝ in. from the edges of the rails to give room for the panel grooves.

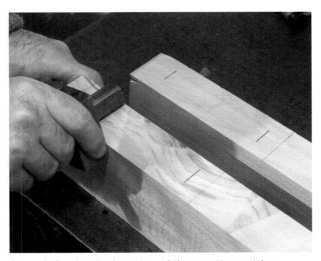

Step 4: Scribe the location of the mortises with a marking gauge, scratching from either side to ensure it's centered.

Step 5: Rout the mortise wall opposite to the fence for the smoothest cut.

Step 6: Chop the center of the ends of the mortises square with a chisel, but leave the corners rounded.

Step 7: Label each joint for reference and mark each edge that gets molded. A dark crayon is most visible, but pencils work too.

My plunge router has a maximum plunge depth of 2¾ in. and my plunge bit is ½ in. by 2½ in. long. I find that these stiles are thick enough to provide good balance and support for the router base. But clamp the two stiles together if you need added support.

It will take two passes for the ½ in. bit to cut the full ¹³⁄₁₆ in. wide mortise. This is a good thing as it will help make a cleaner, more accurate mortise. Set the fence so the bit cuts to the layout line on the opposite side of the stile. Make the first cut to full depth. Do not adjust the fence for the second cut, simply turn the router around and run the fence on the opposite side of the stile. You should end up with a perfectly centered mortise the correct width.

Step 6
Clean up the ends of the mortises: this process begins with ensuring the two routed ends of the mortise are aligned at the end. It's best if both end at your layout line; but if one is longer, re-cut the other the same length to make fitting the tenon easier. I leave the corners rounded from the mortising bit, but trim the widow's peak between them square using a chisel. This allows the tenon to fit tightly against the mortise end without a gap.

Step 7
Lay out the door parts in the orientation that looks best and mark the joints for reference. Label each joint with letters to keep track of which part goes where. Mark each side of each joint with a number or letter to remember how everything goes together. Also mark the edges of the rails and stiles that you will mold.

Step 8
Determine the actual total length of the rails, i.e. the length that shows plus a tenon at either end (even if you've followed these steps to the letter, measurements from the field trump cutlists). Start with 18 in. or

Step 8: Crosscut the rails to length on the tablesawe using a stop block to ensure consistency.

Step 9: Mold the edges of the frame members using a fence to ensure the cut is even over the mortises.

the total width of the door. Subtract the total width of the stiles then add the total depth of the two mortises. Cut the rails to length on the tablesaw. Use a stop block to ensure all three rails are the same length.

Step 9

Cut the quarter round molding on the rails and stiles with a router and fence. For a smooth and straight cut (which is important for the joinery) I don't use a bearing bit because the flat edge of the rails and stiles are interrupted by the mortises. I use a ⅜ in. round-over bit and adjust it to cut a light ⅛ in. deep shoulder. Just remember to mold only the edges you marked, and on both sides.

Step 10

Figure out the length of the tenons, which should be the depth of the mortises minus the width of the molding. If you're annoyed by these recipes rather than just getting the specific dimensions, it's because you've followed the steps perfectly without a nanometer of variation, unlike the vast majority of woodworkers trying to make this door, who, through the accumulation of slight differences at each step, have arrived at a place where specific, planned dimensions

Mitering versus Coping

Why do I miter instead of cope the moldings on this door? And why do I always cope crown moldings? Because that was how I learned and I haven't found a reason to change yet. Seriously. The rationale made sense: on large crown molding miters, the joint will always open up as the wood shrinks, so you should cope as that hides the joint better. On small door moldings, a tight miter will not open up over time, or open just a little (anyway, who's looking?). Using hand tools, mitering takes less time to do and mistakes are far easier to fix. Though coping was done on traditional hand made doors in the past, modern commercial doors are coped because the cope and stick shaper cutters cut them that way. I think the best advantage to coping is that it gives you the opportunity to shout "I can't cope!" in mock despair.

are useless, as I always do. So I tell you how I got them, so you can find your own.

Scribe the tenon shoulders on the rails with a marking gauge. A good way to double check the location of the tenon shoulders is to measure between them (i.e. the face of the rail that will show in the finished door) then add the width of the

Step 10: Scribe the width of the tenons on the rails with a marking gauge.

Step 11: Cut the tenon faces on the tablesaw using a crosscut fence. The rip fence serves as a width stop for consistency.

Step 12: Mark the width of each mortise on its specific tenon.

stiles, minus the molding. If that equals 18 inches, you're good to go. If not, go back and recheck and remeasure until it does.

Step 11

Lay out the thickness of the tenon, scribing it on the ends of the rails. Set a dado blade to the height of the tenons, and cut just shy of your line. You want the tenons to be just a bit too thick for the mortises so you can fit them precisely by hand. Cut the tenons using the tablesaw rip fence as a width stop on the other side of the blade. You can also cut dadoes without a dado blade, using a regular rip or combination blade. It will just take you longer as you have to make more passes.

Step 12

Mark the shoulders of each tenon, matching each to their specific joint. This is key, unless you were so disciplined as to cut only to your layout lines perfectly every time. Cut them on the tablesaw with the dado blade, using the same setup as for the tenon faces, adjusting the blade height for the new depth of cut. Saw out the waste between the tenons on the bottom rail with a handsaw and coping saw.

Step 13

Rout the groove for the panels on the same edges as the moldings. The groove should be ½ in. deep, while the width depends on the thickness of the material you're using and the type of raised panel bit. As I had a nice 6/4 board handy, I used it, but an 8/4 thick board for the panel would be more traditional (the field of the panel would end up being flush with the face of the door).

To calculate the width of the groove: subtract (twice) the height of the raised panel bit from the thickness of your board. My panel bit cuts ⅜ in. deep, so I subtracted ¾ in. from the 6/4 board which I'd milled to 1¼ in., leaving me with a ¾ in. thick tongue on the panel and the same width for the groove.

Step 13: Rout the groove for the panel on the rails and stiles with a slot cutter and fence.

Step 14: Trim the molding off the stiles around the mortises with a handsaw.

Cut the groove close to, but not through the mortises on the stiles. Square up the groove ends with a chisel. The panels are rectangular, and the grooving bit leaves a rounded corner. Use a chisel to cut the groove corners square allowing the panel to fit fully into the groove. Separation between the panel groove and the mortises will ensure glue doesn't squeeze out of the joint and connect the panel to the frame. The panel, by the way, needs to move freely within the frame—go back to those first few chapters if you need a reminder why.

Fitting Joinery

Time to fit the tenons in the mortises. The strength of the joint depends on the fit of the tenon in the mortise. It should not be loose, but not too tight either. You should be able to lightly tap the joints together with a mallet, or even press it together by hand if you're young and strong. To make the joint look good, with tight seams between the faces and the mitered moldings, is the other main goal.

Step 14

Saw away the molding on the stiles where the rails will fit. Saw close but not directly on the shoulder. Do not saw close to your

Step 14: Pare the waste around the mortise with a wide chisel. The face of the mortise should be square and even.

layout lines for the mitered corners. Those get cut later with a chisel. A bandsaw also works well to cut away the waste.

Pare the face of the mortises flush with a sharp wide chisel, making sure you don't cut the shoulder edge itself—use it as a reference to guide your chisel. Use a square to check the mortise face, that it's flat and square, even a little undercut compared to the shoulders. This will ensure the rail and tenon shoulder fit tightly against the mortise face.

MITER GUIDE

A simple jig guides the chisel to cut 45 degree miters on moldings.

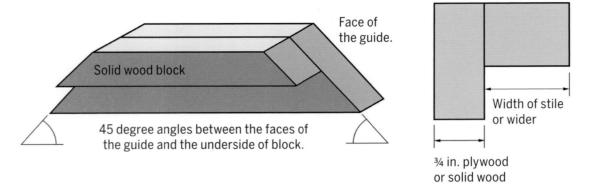

Face of the guide.

Solid wood block

45 degree angles between the faces of the guide and the underside of block.

Width of stile or wider

¾ in. plywood or solid wood

Miter Guide

A simple shop-made jig is essential to mitering the moldings on the rails and stiles. It guides the back of a chisel ensuring it pares the miter face straight, flat and true. The jig is simple to make. Screw or glue a 5 in. by 10 in. scrap of ¾ in. plywood to a straight and square block of solid wood about the same length, 2 in. wide and at least 1½ in. thick (you can use an offcut from the end of one of the stiles or rails if you have it). On a chopsaw or tablesaw miter both ends at 90 degrees (mind any screws). And the jig is done (or up).

Step 15: With the miter guide clamped to the stile, miter the faces of the moldings with a wide chisel.

Step 15

Use a chisel and a shop-made guide to miter the faces of the molded stiles. Chop and pare the 45-degree surface just to just inside of the layout lines you made in **Step 2** (to help orient yourself, imagine the rail and stile intersecting and what should go where). To use the jig, clamp it to the rail or stile and align the shoulder with the layout line on the stiles. By the way, don't miter the molding ends on the rails just yet.

Step 16

To fit the tenons to the mortises, the object is to create two perfectly complementary shapes, one positive (the tenon) and the other negative (the mortise). It's easier to shave the tenons down in the right places than dig more out of the mortises. So consider your mortises done no matter what. Your job is simply to find out how to cut the tenons to fit.

To do this, gauge the amount of material already removed from the tenon against the thickness of the mortise wall by aligning the tenon from the outside. You'll need to rotate the rail 180 degrees so you check the matching side of the tenon against the mortise wall on each side. Look at how the shoulder of the tenon aligns with the face of the mortise wall to see where the tenon is too thick.

Step 16: Test the thickness of the tenons against the mortise walls by aligning the matching faces.

Step 17: Shave down the tenons to match the mortises with a shoulder plane that allows you to cut right up to the shoulder.

Step 17: Round the ends of the tenons with a rough file to match the mortise ends.

Step 17: Pare the waste flat between the tenons and anywhere else on the tenon shoulders.

Step 17

Shave the tenon faces down with a shoulder plane. This kind of plane allows you to cut right up to the shoulder (ha—now you know where the name of the plane comes from) of the tenon. Be careful so you don't create a wedge shape that will split the mortise.

I make pencil marks on the tenon face to keep track of where I have cut and how much. Round the corners of the tenons to fit the rounded ends of the mortises. Pare the shoulders of the tenons flat. Test fit the tenons in the mortises often. The tenon should fit with only light taps from your mallet.

Step 18: Test fit the joints before mitering the rails to ensure they line up with the layout lines.

Step 18: Pare the molding miters to the tenon shoulder, using the miter guide.

Step 19: Dry clamp each joint together to get a realistic sense of what the fit will be like in the finished door.

Step 18

With the joint well-fit, check the intersection of the square end of the molding on the rail against the molding on the stile. If the moldings don't intersect on the layout line, adjust the tenon ends accordingly, sliding the rail one way or the other. Then miter the molding on the rails to fit. This process is a bit harder for the middle rail, as you have moldings on either side to fit, and they have to come together at the same time. Patience, care (and practice) will get you there. As a rule, cut off less than you think and then test fit.

Step 19

If at this point your joints aren't coming together perfectly, well, that's fine and to be expected. None of the processes I've shown you depends on machine-repeatable steps with CNC accuracy. Consider, though, that the football is on the 2 yard line. Now, take the time and effort to get it into the end zone. You have as many downs as you like—a roundabout way to say fine tune the joint surfaces until everything fits like a glove. If you have gaps, realize that where the joint touches describes the places where material should be removed. Then the gaps elsewhere disappear.

To check your progress, clamp each dry joint together. You'd be surprised how many gaps simply disappear when under a little pressure from a clamp.

When you have fitted all the joints individually, fit the whole frame together. You may find that joints that were tight on their own now are open because the other joints pull them in an unexpected way. Sleuthing the right adjustments is a process nearly impossible to put into words. I can only offer advice: pare little, check often. My most common mistake is to solve one adjustment problem and create another by not thinking through all the consequences of my cut.

And there is always hope: a simple mixture of sawdust and PVA glue makes a good fill for gaps in molding and joints.

Step 19: Dry clamp the whole door when you've fitted all the joints together to see how the joints fit as a whole.

Epoxy Fill Technique

Lots of cracks in a board certainly weaken it, but doesn't mean the wood is unusable. You can let them be in a panel. Or they can become a "feature" with the help of structural epoxy—just fill the cracks with it.

If the cracks go through the board, seal the cracks on the back and edges with painter's blue tape. Any tape will likely work, but the painter's tape is easiest to take off. Flip the board over and set it on blocks above your workbench. This helps to locate any leaks and avoid gluing the board to the bench. Mix West System 105 resin and 207 hardener and pour into the cracks. Replenish as the epoxy works its way down into the cracks. Bubbles may form at the top—break them and add more epoxy. If you find voids, add epoxy again— new epoxy blends perfectly with old. If any leaks develop (check under the board often), seal them with more tape. When the epoxy dries, sand it evenly with the surface of the board and prep the surface for finishing. This treatment works well for cracked frame members as the epoxy adds strength. Apply finish normally.

Seal the cracks on one side with painter's tape.

Fill the cracks with a sandable structural epoxy.

Sand the dry epoxy flush with the board's surface.

Step 20: Measure for the panels between moldings with the door dry fit together.

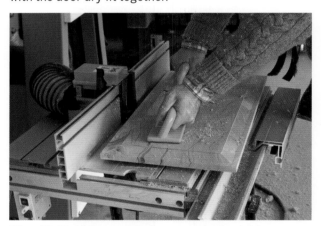

Step 21: Mold the panel edges on a router table, adjusting the cutter height to leave a tongue the width of the groove you've cut in the rails and stiles.

Step 23: Hand sand the moldings before assembly to 320 grit for a satin sheen.

Step 20

Measure the real interior dimensions to get your panel sizes. If you have a ½ in. deep groove, add ¹⁵⁄₁₆ in. to the length and width of the panels. If you're working at a dry time and expect the door will experience greater humidity later, I'd add a little less to the width of the panel, perhaps ¹³⁄₁₆ in. or ¾ in.

Step 21

Cut the panels to size and mold both sides of each one. The 2¾ in. panel raising bit I use fits my router table just fine. It's a very big bit, so I take multiple passes cutting just a bit at a time. Of course, set the bit height to leave a tongue on the panel that easily but snugly fits the grooves in the frame.

Step 22

This step is key to avoid glue up disasters: dry fit your entire door, panels included. Now is the time to find out if your panels are slightly too large, or you forgot to chop out the corners of the grooves and other details, before there's setting glue to complicate matters.

Step 23

Sand the moldings on the frame members before assembly. They are much easier to sand when the frame is apart. I use a sanding block to get into the tight shoulder area, but do not sand over the joint surfaces. I hand sand the rounded molding surface. With this door, I will sand up to 320 grit because it gives a better satin shine with the oil and wax finish.

Step 24

Sand the panels to 220 or 320 grit. Then finish them before assembly. Finishing the panel before assembly has several advantages. First, it will help prevent glue squeeze out from the joinery attaching the panel to the groove—remember that the panel needs to float freely in the frame (though not rattle around in it). The second reason is when the panel shrinks, it won't reveal an unfinished edge. I used several coats of Danish oil then a wax buffing. The grain is wild and rustic, so the oil brings that out. It's an interior closet door so I'm not too worried about protection.

Step 25

Glue up can be with any glue you like, but I did this one with yellow glue. Pour a goodly amount of glue into a cup or bowl and thin it with about 5% water (just a few drops). Regular yellow glue has a fast tack and I want to buy a little time to adjust the joints. The added water doesn't weaken the bond, but does slow curing.

Work the glue on all the surfaces of the mortises first (in the superstitious theory that they will dry out more slowly inside), then on all the surfaces of the tenons.

Tap the tenons into all the mortises of one stile and adjust the joints as best you can. With the door on edge, slide the panels into place. Finally, fit the other stile on top and clamp the frame tight. Set the door on edge overnight to let the glue dry. If you lay the door flat, make sure it is evenly supported so it doesn't dry with a curve in the face.

Step 24: Apply a coat of oil to the panels before assembly.

Step 25: Apply slightly thinned yellow glue to all tenons and mortises before assembling the parts.

Step 25: Slide the panels into the half-assembled frame, letting gravity help.

Step 26: Drill ¼ in. diameter pilot holes for the ¼ in. square pegs, centered on each mortise.

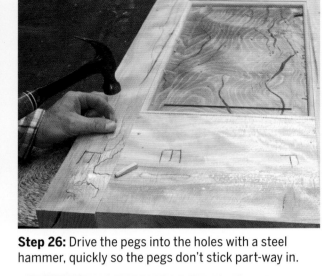

Step 26: Drive the pegs into the holes with a steel hammer, quickly so the pegs don't stick part-way in.

Step 26: Trim the pegs flush with a veneer saw or a flexible handsaw, taking care not to scratch the surface.

Step 27: Sand the rails and stiles flat and smooth to 320 grit without touching the panels.

Step 26

Make peg stock from white oak if you have it. If you don't, any of the harder woods will work (red oak, locust, maple even walnut). Rip the edge of a wider board so the offcut is ¼ in. by ¼. Crosscut the peg stock into individual pegs about 1¾ in. long then taper the ends. I use a chisel for tapering (keeping fingers out of the way), but a sander works as well (again, fingers).

When the glue has dried in the door, lay out and drill for the pegs. They should be centered in each tenon (you remember the dimensions and location of your mortises,

right?) If your mortises were 2½ in. deep before you cut the moldings, then they're about 2⅛ in deep finished, so center the peg about 1¹⁄₁₆ in. from the joint edge.

If you want to see the pegs, drill ¼ in. holes on the front of the door nearly through to the other side. Put the pegs on the back if you don't want to see them.

With a drop of yellow glue in the peg hole, use a steel hammer to pound them in (support the door well on the back side so the peg doesn't break through). Trim the proud peg with a veneer saw or other flexible backless saw, or use a chisel to pare them flush.

Step 27

Sand the rails and stiles on both sides of the door to 320 grit for a satin look with an oil finish. Take care not to sand the panels. One advantage of using a thinner panel is that it's recessed, making sanding the door easier.

Step 28

Trim the ends of the door with a track saw and guide. You can also do this work with a circular saw, but the saw's base can scratch the door surface and its hard to get accuracy without a jig.

Step 29

Finish the door faces with Danish oil. After the first coat dries, use 320 wet-dry sandpaper to sand in a second and a third coat, each time being sure to wipe away the excess. The dust-oil slurry fills the pores just a tiny bit and adds a bit to the sheen when dry. I don't bother finishing the edges yet as the door has to be hung. In fact, you can wait until you've hung the door to finish the front and back as well (but that means refitting all the hardware). But certainly wait until you've fit the door to the jamb to finish the edges.

Congratulations—you're now the proud parent of a door slab.

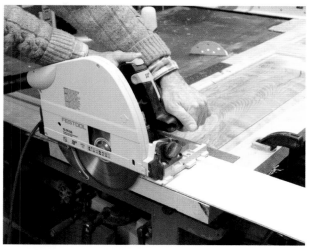

Step 28: Trim the door ends square and to length with a track saw and guide.

Step 29: Finish both faces with several coats of oil for a satin sheen.

Larger doors can be
made from plywood
pieces fit together.

Construction-Grade Plywood Door

8

Who says a good interior door has to be fancy, complicated or take a long time to make? A good friend and life-long contractor, Glen Hochstetter, came up with a reasonably durable pocket door that's about as fast, cheap and easy to make as a door can be, if you like the inescapably modern look. The design celebrates humble materials in a perfectly functional, minimalist look.

The structure is a simple torsion box. The core is a frame of 5/4 by 4 in. common #2 pine held together with pocket screws. The exterior and edges are ⅜ in. thick AC plywood glued and nailed to the frame. The edge treatment is strips of plywood to make the door look as if it's solid plywood. The door has reasonable stability and pretty good durability. And if the edges of the plywood delaminate, it is just part of the look. It is the carpenter's jobsite version of the honeycomb core door in **Chapter 10**.

Variations

This design can accommodate a lot of variation and substitutions without affecting performance. The interior frame can be thicker or thinner, though not much thinner than 5/4 stock pine for a standard 36 by 80 in. door. The plywood can be any variety as long as it is either veneer or lumber core (using particleboard or MDF core plywood

would substantially weaken the design and increase the weight). The plywood skin does not need to be a single piece, but can be a mosaic of pieces (just ensure you add frame members underneath all edges). And these scraps don't need to be the same thickness. A door mixing thicknesses could be quite beautiful, though more difficult to make flat.

Plywood edges do not take screws well, so the frame at the top and bottom is exposed (no one sees it anyway). Surface mount hinges such as strap or H hinges will work fine, as will a pocket door hanging track or a pivot hinge. Butt hinges, or any hardware that needs to screw into the edge of the plywood, is not advisable. For a butt-hinged version of this door, set the frame closer to the edge, perhaps ¼ to ½ in. deep, where hinge screws can bite into the solid pine.

If you'd like to reduce the sound and heat transference through the door, pack the interior of the frame with 1 in. thick rigid insulation.

Parts and Assembly

Step 1

At the lumberyard, pick common pine boards that are as straight as possible. Sight the boards along the length to make sure. Pick sheets of plywood with as little warping as possible.

CONSTRUCTION-GRADE PLYWOOD DOOR

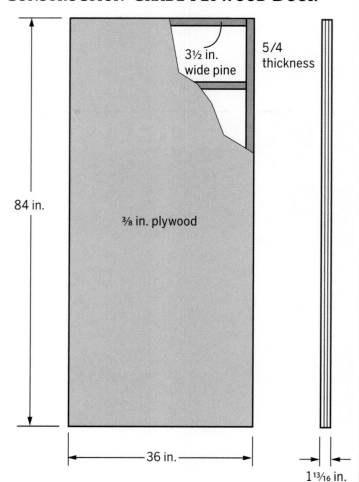

84 in.

3½ in.
wide pine

5/4
thickness

⅜ in. plywood

36 in.

1¹³⁄₁₆ in.

FINISHED DIMENSIONS:
36 in. by 84 in. by 1¹³⁄₁₆ in. thick

MATERIALS
- 2 @ 3½ in. wide, 5/4 common #2 pine, 8 ft. long
- 1 @ 3½ in. wide, 5/4 common #2 pine, 12 ft. long
- 2 @ 4 ft. by 8 ft. by ⅜ in. thick AC plywood

CUTLIST
- Stiles 2 @ 3½ in. by 84½ in. by 1⅛ in. pine
- Rails 4 @ 3½ in. by 27 in. by 1⅛ in. pine
- Skins 2 @ 36½ in. by 84½ in. by ⅜ in AC plywood
- Edging strips 8 @ 1¼ in. by 84½ in. by ⅜ in AC plywood

HARDWARE
- Hanging door track

Step 1: Sight down the pine stock to pick straight and true boards for your stiles and rails.

Step 2
Cut the four pine rails to 27 in. long, perfectly square at both ends. Cut the stiles to 84½ in. long. You want a finished frame that's 34 in. wide by 84½ in tall. The frame is not as wide as the finished door to accommodate the plywood strips on the edges. The frame should be flush with the top and bottom of the finished door.

Step 3
Set up a pocket screw jig for 1⅛ in. thick stock, locking the depth collar in the correct position. If you prefer, you can join the frame together with biscuits, floating tenons or any other way you like (the strength is in the plywood skins glued to the frame).

Step 4

Drill two pilot holes in both sides of both ends of each rail. When you drill holes on the opposite side, offset the rail by ½ in. so the holes don't line up with the ones on the other side. Pocket screws set from just one side can sometimes twist the frame members. Remember, they're fresh from the lumber yard and may have a little twist or warp in them (though you did your best to minimize it). Adding screws from both sides helps keep the parts in alignment.

Step 5

Space the four rails evenly from top to bottom. Screw them to the stiles using coarse thread 1½ in. long pocket hole screws. Screwing the frame together on a flat bench will help keep the frame members in alignment. If you don't have a large bench, put down plastic sheeting and work on the floor. Clamp the frame members together to keep them in alignment if necessary.

Step 6

Rip the plywood sheets to 36½ in. wide with a circular saw and guide (I don't recommend ripping full sheets of plywood on a tablesaw unless that tablesaw is a slider). Remember that you can choose which part of the sheet you want to use, looking at grain and character, and rip your door fronts from one side, the other side, or from the middle of the sheet.

Step 7

Crosscut the front and back to 84½ in. long, again using a circular saw and guide.

Step 8

Spread glue on one side of the frame. A brush spreads it evenly, but a shingle can be faster if you're working alone.

Step 4: Drill two pocket screw holes on both sides and ends of each rail.

Step 5: Screw the frame togther with 1½ in. pocket screws.

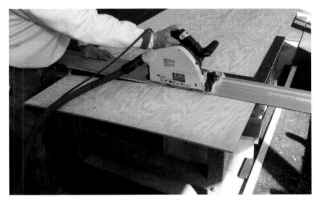

Step 7: Crosscut the plywood fronts and backs to 84½ in. long with a plunge saw and track.

Step 8: Apply glue to one side of the frame with a brush.

Step 9: Align the plywood front flush with the top of the frame and 1¼ in. in from the sides.

Step 10: Use your body weight to clamp the door together while nailing the plywood to the frames.

Step 9

Align one sheet on top of the frame so it hangs 1¼ in. off both sides and is aligned with the bottom edge. If you're working alone, drive a small nail in one corner to enable you to align both ends at the same time. It works like a pivot so the whole sheet doesn't slide around. Leave the nail proud so you can get it back out if the sheet is out of alignment.

Step 10

Nail the sheet down to the frame with a pneumatic gun using 18 gauge 1½ in. long nails. They're large enough to offer good holding power, yet small enough that the holes are hard to spot. Lean or kneel into each shot as you want the plywood and fame to be in contact. You can also nail the sheet down by hand with 4d finish nails. Flip the assembly over and repeat the process for the second sheet on the back. Nail the plywood to internal rails to give some clamping pressure across the face. Don't shoot nails too close to the edge as you will trim the door to size later.

Step 9: Drive a small nail into one corner to keep the sheet from moving around as you align it at the other end.

Step 11: Rip eight plywood strips 1¼ in. wide using a push stick. These are to fill the edges of the door.

Step 12: Rip insert pieces thinner on the tablesaw, halfway from each side so your hands don't go near the blade.

Step 11

Rip eight strips 1¼ in. wide from the plywood offcuts on the tablesaw. You'll use these to fill the edges to make the door look like it's solid plywood.

Step 12

Fit the strips into the slots on one edge, four for each side. It may be that three strips fit neatly if the plywood is on the thicker side or your ¾ pine is on the thinner side.

Step 13: Spread glue on all faces of the strips and on the inside of the slot.

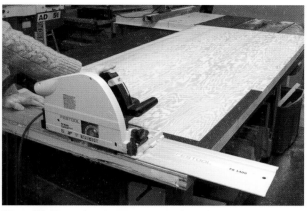

Step 15: Trim the door to size with a plunge saw and track.

If not, rip two of the strips a little thinner and put them between the full strips. Test the fit of the four in the slot on both sides— the fit should be snug, but not so tight.

Step 13

Spread glue on all sides of the strips and on the walls of the slot. Sandwich them together and set them into the slot. If there are gaps, clamp along the length. You don't want to use nails along the edges as you'll later cut the door to width. After the glue sets, repeat the same process for the other side.

Step 14

If you leave the door overnight to cure, stand it up on edge with air circulation all around it. Leaving it flat on a bench or the floor will encourage warping, as moisture release/absorption will be uneven.

Step 15

When the glue has cured, rip and cross cut the door to its final dimensions with a plunge or circular saw and guide. Use a framing square to make sure you create a rectangle, not a parallelogram.

Step 16

The surface of AC plywood is relatively rough and splintery, thereby difficult to sand smooth. If you want to remove dirt and layout lines, sand carefully by hand with 180 grit, trying

Step 17: A finger slot to open and close the door can be added with a bullnose router bit.

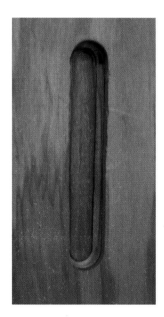

not to catch the grain. Sand the edges and corners with a block and 120 grit, grinding them down such that you can't catch your nail on the surface veneers.

Step 17

Rout a simple 4 in. long finger pull on either side about 36 in. from the floor. Use a ¾ in. bullnose bit and a router with a fence as a guide. Or add handle hardware if you prefer.

Step 18

Leave unfinished for an industrial look. Or finish with oil. Film finishes will add to the rough look. Realize that the coarse surface will not let a film finish flow out easily, and sanding it would be difficult.

Congratulations—you're now the proud parent of a door slab.

Modern Rustic Door

9

So we're back to a rustic look much the same as the board-and-batten door, though no nails show through the front. The difference is this one has a true mortise-and-tenon frame core. Unlike a real board-and-batten design, this door is durable, has a good R-value and is exterior-grade. The downside is that it's harder to make, uses more materials and is heavy. Rather unbelievably heavy. So heavy that if you're not a husky teenager, I recommend hiring one to move the completed door around your shop. Whew.

The durability in this door is in the frame: the boards on the front and copper cladding on the back are largely decorative and insulating. This frame construction is as simple as it gets, making it a good place to start if you've not done one like it before, simpler than **Chapter 7**. There are no moldings and the entire frame gets hidden, giving you the freedom to make as many mistakes as you like.

Variations

The copper cladding on this door was a specialty item and was applied by a professional. It involves soldering and applying a faux patina. If you'd prefer not to expand your skillset this way, you've got a lot of alternatives for finishing the inside.

The first and easiest is to attach tongue-and-groove boards to both sides. If you do this, it's

a good idea to use thinner exterior boards and a heavier frame to support them, though the balance depends on the weight of the wood. For a heavy door such as oak, I'd make the exterior boards about ⅝ in. to ¾ in. thick on both sides and the interior frame about 1 in. to 1¼ in. thick. With light woods such as butternut and pine, you can make the frame the same thickness as the exterior boards, i.e. all ¹³⁄₁₆ in. This is easier to do, frankly, as you can buy wood all the same thickness and mill it all at the same time.

Attaching the exterior boards without visible fasteners is possible on one side (as in this chapter), but not so easy on both sides. You can nail through the tongues for the boards in the middle, but you can't do that on the edges. Gluing just the edges of the door is a good solution, but also remember that a few finish nails through the face can add to the rustic look.

Another option is to simply finish the inside of the frame by choosing cabinet grade plywood for the panels and applying a bolection molding, as in **Chapter 11**.

Step 1

Mill your rails and stiles following the steps in Chapter 2 for Rough Milling and Fine-Tune Milling, referring to Steps 1–2 of Chapter 7 as well. The specific widths of the rails are not too important, nor are their locations,

MODERN RUSTIC DOOR

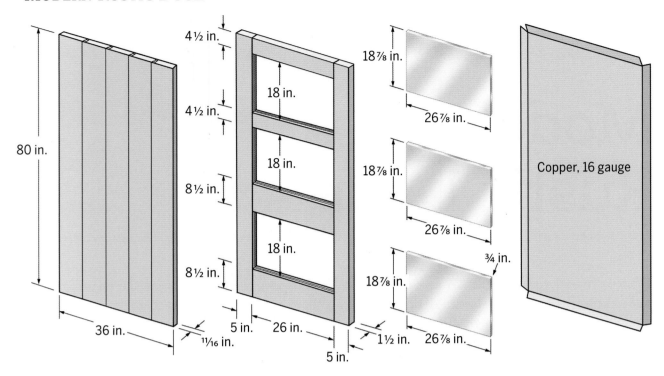

FINISHED DIMENSIONS:
36 in. by 80 in. by 2¼ in.

MATERIALS

- Approximately 25 board feet of 8/4 African mahogany for the rails and stiles
- Approximately 24 board feet of 4/4 white oak
- 1 sheet of exterior grade plywood
- Rigid insulation
- Copper sheeting ¹⁄₁₆ in. thick
- Scrap of white oak for the pegs
- Approximately 25 exterior grade screws, 1¾ in. long
- Approximately 25 exterior grade screws, 1¼ in. long

CUTLIST

- Stiles 2 @ 80 in. by 5 in. by 1½ in. African mahogany
- Rail 1 @ 31 in. by 8½ in. by 1½ in. African mahogany
- Rail 1 @ 31 in. by 8½ in. by 1½ in. African mahogany
- Rail 1 @ 31 in. by 4¼ in. by 1½ in. African mahogany
- Rail 1 @ 31 in. by 4¼ in. by 1½ in. African mahogany
- Panels 3 @ 18⅞ in. by 26⅞ in. exterior grade plywood
- Planks 5 @ 80 in. by 7⅝ in. by ¹¹⁄₁₆ in. white oak
- Pegs 12 @ 1¾ in. by ¼ in. by ¼ in. white oak

HARDWARE

- Three 5 in. butt hinges
- Knocker
- Lockset
- Deadbolt

as they will be covered with copper on the back and covered by boards on the front. As I had 9 in. wide boards, I made the two lower rails 8½ in. wide, and the upper ones 4½ in. wide. This configuration might look odd if the frame was visible, but it isn't.

Step 2

Lay out the door parts and mark the joints for reference. Label each joint with letters to keep track of which part goes where. Clamp the stiles together and lay out the mortises on both at the same time. Mark where the rails intersect the stiles.

Step 3

Mark the length of the mortises in the stiles. The mortises should be ⅝ in. in from either side of the rest of the mortises to allow for the plywood panel grooves. At the ends of the stiles, they should be at least 1 in. from the top and bottom of the stiles.

Step 4

Mark the width of the mortises with the marking gauge. Scribe from both sides to ensure you've centered the mortise. Make the mortise a little more than half the thickness of the stile, about ⅝ in. or 9/16 in.

Step 5

Cut the mortises, following **Step 5–7** in **Chapter 7.** As the edges of this door will be covered in copper, you can (if you want a really bullet-proof door) mortise the stiles from both sides and make through-tenons. These do not have to be wedged, but simple through joints and they would be stronger. Still, simple 2½ in. deep mortises work well.

Step 6

Determine the total length of the rails by subtracting the width of both stiles from 36 in. and adding the depths of the mortises (even if you've followed these steps

to the letter, measurements from the field trump all calculations). Cut the rails to length on the tablesaw. Use a stop block to ensure all four rails are the same length.

Step 7

Scribe the tenon shoulders on the rails with a marking gauge. This measurement should be just a little less than the depth of the mortises. You will end up with a door slightly wider than 36 in., but this gives you wiggle room to trim down after assembly.

Step 8

Scribe and cut the tenons on the tablesaw with a dado blade, following the specific instructions in **Steps 11–12** in **Chapter 7.**

Step 9

Fit the tenons into their respective mortises. Take your time and get them perfectly flush, as the fit will affect the glue joint between the two door sides. Refer to **Steps 14, 16–17** in **Chapter 7** for a more detailed description of fitting the tenons. These are easier as there's no molding to miter.

Step 10

When you have fitted all the joints individually, fit the whole frame together and adjust the joints as necessary. Remember, they get hidden, so accuracy is more important than beauty.

Step 10: Dry fit the door together to test the fit of the joints.

Step 11

With the door still dry fit, use a router and grooving bit to cut a ⅜ in. wide by ½ in. deep groove for the panels. If you don't have a grooving bit to do this work, you can also use the dado blade on the tablesaw. The difficulty here is that you should not run the dado through the mortises, but start and stop the groove around them. This can be difficult as you can't see the blade during the cut on the tablesaw. You can mark the front and rear of the blade on the tablesaw surface to act as a guide.

Step 12

Measure from the bottom of the grooves to size each plywood panel. Cut them about ¹⁄₁₆ in. smaller than the full depth of the grooves to ensure they don't prevent the door from coming together completely.

Step 13

Cut a ⅝ in. by ⅜ in. rabbet on the edges of the panels using a dado blade on the tablesaw. You can make the rabbet only ½ in. wide, but I like a little gap around the edges to ensure the panel doesn't keep the frame from coming together. Fit the panels into the dry fit frame. The panels should fit snugly in the grooves, but not require a mallet to set them. Their faces should be flush with the front of the door. Dry fit the entire door. Getting it apart may not be easy, but this step is necessary to ensure the door will go together before you have glue in the joints.

Step 13

Glue up the frame and panels using a slow set, low tack waterproof glue such as polyurethane or an epoxy. Add a bead of glue to the panel grooves, as this will add to the structural integrity of the door. By the way, never do this with solid wood panels.

Step 14

After the glue has cured, rough sand both sides, evening the faces of the joints and ensuring the panels are flush with the frame. Peg the joints, following the details described in **Step 26** in **Chapter 7**. Use a durable wood such as white oak or African mahogany for the pegs and remember that you won't see them when the door is done.

Step 15

Mill five white oak boards for the door front 4 @ 7½ in. wide and one (an edge board) at 7⅛ in. wide, all five a few inches over 80 in. long. There's substantial math going on in these widths, as you need to leave extra for the ⅜ in. tongue on four of the boards. The thickness can be anywhere from ¾ in. to ½ in. Much thicker and you add a lot of unnecessary weight. Much thinner, and you can't hold the boards to the frame with screws from behind. The ¹¹⁄₁₆ in. in this project was necessary for the door thickness to meet the specifications of the jamb.

PANEL GROOVE IN FRAME

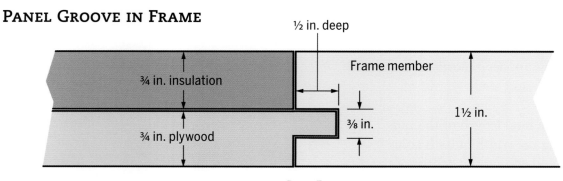

½ in. deep

¾ in. insulation

Frame member

¾ in. plywood

⅜ in.

1½ in.

Door Front

Step 13: Dry fit the entire door to ensure the panels fit and are flush with the frame.

Step 13: Add a bead of glue to the grooves during glue up to make the plywood panels structural members.

Step 14: Hammer oak pegs through all the joints.

Step 17: Rout the grooves in the oak boards with a ³⁄₁₆ in. wide bit. Cut from both sides to center the groove.

Step 16

Lay out the boards in the orientation you want, but put the 7⅛ in. wide board on one edge. Number them from left to right and mark the edges for whether they stay square (outside edges of first and last boards), have a tongue, or have a groove (to be clear, the 7⅛ in. wide board gets a groove). This step is really key to avoid going back to the lumberyard for more white oak.

Step 17

Cut centered grooves about ¼ in. wide and ⁷⁄₁₆ in. deep on the right boards using a router and grooving bit. It's best to use a ³⁄₁₆ in. wide bit and flip the board for a second pass to ensure the groove is centered. Alternately, this can be done on the tablesaw using a regular combination blade running the board up on its edge.

Step 18: Use a dado blade on the tablesaw to cut the centered tongues on the oak board edges.

Step 18

Cut the ¼ in. thick and ⅜ in. wide tongues on the appropriate boards. They should be a little shorter than the groove is deep so you can get a good tight seam on the front. Make a few test cuts to ensure the tongues fit snugly in the grooves, but not so tight that you have to use a mallet to set them. I use a dado blade for this work, but a regular combination blade works too, though you have to make a few passes.

Step 19: Lightly chamfer the tongues before assembly to help the boards fit together easily.

Step 21: Drive exterior grade screws through the panels and frame to attach the oak boards.

Step 22: Trim the oak boards flush with the frame using a plunge saw and track.

Step 19

Chamfer the edges of the tongues with a block plane to help them fit. You can also chamfer the outside edges if you want that look to hide the seams, as described in **Step 7** in **Chapter 6**.

Step 20

Clamp the five boards together, tongues snug in grooves, and put them face down on the bench (they should be about 36⅛ in. wide total. Apply waterproof glue to the faces of the boards at either side, but only 1 to 2 in. in from the edges. Clamp the door frame to the boards. You might note that you can only clamp along the edges. This is fine.

Step 21

While, or after, the glue sets, drill pilot holes and set screws through the frame into the backs of the oak boards. It's important to capture only one side of each board. The outside board edges are glued, so set screws along the edges of the board next to it. Use 1 ¾ in. long screws through the frame and 1¼ in. long screws though the panels. Any type of exterior grade screw will work. I prefer the stainless steel screws in spite of their expense from the superstition that they bend easily with wood movement and won't cause cracks as often. When you set the screws in the middle of the door, get up on top of it. Your weight will help clamp together the boards and the frame, ensuring a flat door front when you're done. By the way, stainless steel screws are very soft, so drill good pilot holes and back the cam off on your cordless driver, or you'll have broken heads to drill out.

Step 22

Trim the edges and ends with a circular saw and guide.

Step 23

Cut three ½ in. thick plywood panels sized to fit inside the frame on the inside of the door. Use a waterproof glue and 1¼ in. long exterior grade screws to set them in the frame. This will make the panels flush with the frame, and ready for the application of the copper or more tongue and groove panels.

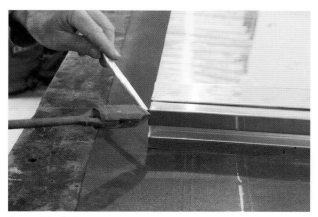

Step 28: Solder the corners to keep the edges secure.

Step 29: Overlay fake copper strap hinges to hide the seams on the face.

Congratulations—you're now the proud parent of a door slab except for the copper.

Copper Cladding

Copper cladding a door properly does take some professional tools, such as a metal brake, but you can still do a good job without them (if you have the proper tools, then you know what you're doing and don't listen to me). I'll walk you through the basics of doing it in a wood shop.

Step 24

After you've hung the door and made sure all four edges fit the jamb and sill just right, take the door down and remove the hardware. With a track saw, cut a ¹⁄₁₆ in. wide rabbet around the edges, about half the thickness of the door. The copper sheet needs this rabbet to fit in, so there's no protruding sharp edge around the perimeter.

Step 25

Using .021 in. thickness copper, cut pieces to size with either a pair of metal shears or a jigsaw with a metal cutting blade (slow the speed down). The shears will leave a puckered edge while the jigsaw will leave a flat edge. I think both look fine. As most copper sheets are smaller than a full-sized door, place your seams in places that will be easy to hide with pretend strap hinges. Cut squares out of the

corners to make folding the sheets over the edges and seaming the corners easier. Make cut outs around the hinges and hardware.

Step 26

Without a metal brake, creasing the edges is difficult but not impossible. Place the copper sheet on the door aligned with the right overhang along the edges. Clamp a thick board over the copper and aligned with the door edge. Use a second board to put even pressure across the face of the copper and bend it over as far as you can with hand pressure. Finally, use clamps to secure the second board to the edge of the door, creasing the copper at 90 degrees. Do this for all sides.

Step 27

Secure the sheets on the door with a very thin layer of construction adhesive, pressing evenly across the surface and letting it cure.

Step 28

Solder the corners and file them down when they cool.

Step 29

Make fake strap hinges out of the copper, folding the edges so they're not sharp to overlay and hide the hinges. Secure them with copper nails.

Butternut is light enough to make the interior frame the same thickness as the panels on the outside.

Honeycomb Core Door

10

Have you ever heard the term "hollow core door" and shuddered? As a snooty custom cabinetmaker, I have. I equate them with mass-manufactured cheapness: the kind of doors that weigh 6 ounces, can be punched through by an 80-lb. nerd without breaking a sweat, and have the veneer edges peeling off before you look at it. Well, I'm not going to show you how to make one of those. Even if you think a hollow-core door is a perfectly nice thing. Instead I will show you how to use the same technology to make a strong, lightweight interior door that we can all agree is nice.

This door uses a resin impregnated honeycomb (essentially a cardboard lattice) to fill the interior. It produces an extraordinarily light, rigid and strong door. This is torsion box construction, a construction approach that can be used for a wide variety of woodworking projects where light, rigid panels are useful, from desktops to long shelves to large cabinets. The major improvement of this design over commercial doors is the addition of solid wood edging. This is not that difficult to do but results in a far more durable door in which the veneer edges are nearly impossible to chip or lift.

The barrier to this kind of construction is the equipment necessary to make it—I use a large veneer press to bond the plywood panels to the core. However, the veneer press is not strictly necessary (see **Variations** below). An additional advantage of the press is that you can then veneer the door itself—Amboyna burl anyone? You might ask if you couldn't just start with ¼ in. thick Amboyna burl plywood. However, where I live only ¾ in. thick figured plywood is available, so thin sheets would be an expensive special order.

Finishing this door is another question: the design asks for paint, though it can also honestly be what it is, and there is beauty in that.

Variations

The vacuum press makes this door easy to make; however, you don't need the press to make a version of this door, or any torsion box. Simply nail or screw the plywood skins to the frame as in **Chapter 8.** You can use decorative screws in an orderly pattern, or you can fill the holes with bondo before painting. Another option is to use a number of gently curved clamping cauls to clamp across the surface (clamped on either side). They can be made from 2 x 4's and reused, though I can't speak to the practicability or effectiveness of this approach as I've never done it.

Exterior grade honeycomb core doors are also possible—just substitute marine

HONEYCOMB CORE DOOR

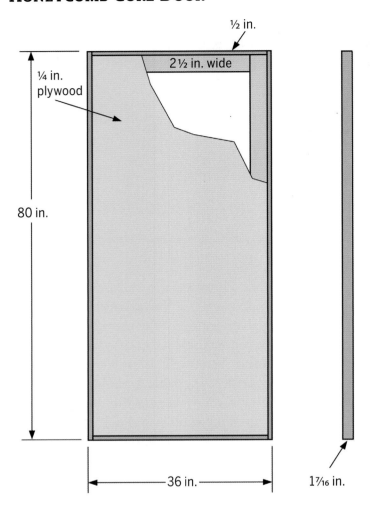

½ in.

2 ½ in. wide

¼ in. plywood

80 in.

36 in.

1⁷⁄₁₆ in.

FINISHED DIMENSIONS:
36 in. by 80 in. by 1⁷⁄₁₆ in.

MATERIALS

- 2 @ 4 ft. by 8 ft. by ¼ in. C-3 grade birch plywood
- 20 square feet of 1 in. thick resin impregnated honeycomb
- 7 board feet of 5/4 poplar
- 2 board feet of 8/4 hard maple

CUTLIST

- Skins 2 @ 35⅝ in. by 79⅝ in. ¼ in. birch plywood
- Stiles 2@ 2½ in. by 79⅝ in. by 1 in. poplar
- Rails 2 @ 2½ in. by 30⅝ in, by 1 in. poplar
- Block 1 @ 2 in. by 20 in. by 1 in. poplar
- Edging 1@ 20 in. by 2 @ 80 in. by 2 in. by ½ in. maple
- Edging 2 @ 26 in. by 2 in. by ½ in. maple

HARDWARE

- Tubular latch set
- Three butt hinges

grade plywood for the surface skins, durable wood for the interior frame and edging and use waterproof glue. This door design also accommodates swinging pet doors better than others. Simply design an internal wood frame around the opening necessary for the door to give strength to the door after the hole has been cut.

Step 1

After buying your materials, you're nearly done, this door is so easy. Just a few more steps. Ha ha.

Step 2

Rip and crosscut your two plywood panels approximately 35⅝ in. by 79⅝ in. You can make them bigger for more wiggle room when you cut the door to size, but not much smaller than this.

Step 3

Mill the poplar rails and stiles precisely 1 in. thick and at least 2½ in. wide. If they are any thinner or thicker, they'll cause problems in the lamination as they need to match the honeycomb material, which is precisely 1 in. thick.

Step 2: Trim both thin sheets of plywood to size at the same time with a plunge saw and track.

Step 4: Align the frame members and mark the joints for dominos.

Step 6: Glue and clamp the frame and block on the lockset side.

Step 7: Fit pieces of the carboard honeycomb sungly together within the frame. They do not need to be glued to the edges.

Step 4

Cut the stiles 79⅝ in. long and the rails 30⅝ in long. Layout your frame and mark your joints for domino tenons. Alternately, #20 biscuits work well as do pocket screws, though I'd be careful using pocket screws here as the frame is not very wide and you'll be trimming it to size later.

Step 5

Add an approximately 20 in. long block to the lock side of the door, centered 36 in. from the bottom. This block will give you solid wood to set a latch or lockset.

Step 6

Clamp up the frame with yellow glue. After the glue cures, scrape or sand any glue excess off the joint faces.

Step 7

Cut the honeycomb panels to fit. The ones I use come in 2 ft. by 4 ft. panels, so I'll use almost exactly two in this door, but in parts and wedged tightly against each other. There is no need to glue them to one another or to the frame.

Step 8: Spread glue evenly over the entire surface of the inside face of the panel using a foam roller. Be sure to spread adequate glue at the edges.

Step 9: Lay the frame onto the panel and carefully center it.

Step 10: Keep the frame and panels aligned while the vacuum takes hold.

Step 8

Lay down plastic on the veneer press platen (I use dry cleaner bags) to avoid filling the vacuum channels with glue or gluing your door to the platen. It's best to tape them down so they don't move out of place. Place either the back panel or the front panel on the platen (with the outside face down, of course. Roll out a reasonably thick layer of glue (I used a urea resin: Pro Glue) on the sheet, making sure to get good coverage at the edges.

Step 9

Set the door frame on top of the plywood and align the corners as best you can. They don't need to be perfect, but get them close.

Step 10

Roll out glue on the inside face of the second plywood panel then align on top of the frame. I find the water content of the glue warps the sheet a bit, so I tape down the corners to keep the parts aligned before closing the vacuum press.

While the vacuum takes hold, you'll have some time to make sure the parts are aligned and don't slide around too much. Urea resin glue is temperature sensitive, so if your shop is below 70 degrees, put an electric blanket over the door to help it cure. I put the pot of left over glue on top of the blanket so I have a gauge of when the glue has cured and can take the door out.

Step 11

Trim the edges and ends of the door square with a plunge saw and track, taking small amounts from either side. In the offcuts, look for delamination. If you find it, check to see how far and how bad it goes, prying with a knife. If the plywood pulls away from the poplar core, work the spot open as much as you can with a putty knife, insert fresh glue and reclamp. Do this anywhere you find delamination at the edges. It can be caused by a number of problems, mostly glue starvation by missing a spot when you rolled the glue out. Finally trim the door to a rectangular 79 in. by 35 in.

Step 11: Check the offcuts from the edges to see if you have any delamination between the frame and plywood panels.

Step 12: Glue the maple edging to the top and bottom of the door first, using long bar clamps.

Step 12: Trim the edging flush with a router and flush trim bit.

Step 14: Sand the maple edging flush with the door face, being extraordinarily careful not to burn through the thin plywood veneer.

Step 12

Mill the maple edging ½ in. thick by 1⅝ in. wide, or just a bit wider than the door is thick. Clamp overlong pieces to both ends first and let the glue cure. Remove the clamps and flush trim the edging with the face of the door. I use a router with a flush trim bit only because the door is thick enough to offer a stable surface for the router base (I would not do this on thinner panels). The right tool for this job is a lipping planer, but I do not own one. Festool also makes a horizontal base for its MFK router that works well for edge trimming thinner panels.

Step 13

Trim the extra length of the edging flush with the sides. Then apply the long edging to the sides, repeating the process as you did for the ends.

Step 14

Sand the edging and door faces carefully, remembering that it is easy to sand through commercial veneers as they are extraordinarily thin.

Congratulations—you're now the proud parent of a door slab.

Frame-and-Panel Two-Face Door

11

In the architectural millwork shop where I apprenticed, we would have a curious problem. We'd panel one room in mahogany and the one next to it in butternut. In what wood should we make the door between the two rooms? The answer was both woods—in the mahogany room it was a mahogany door, and in the butternut room it was a butternut door. This amazed me at the time, and still does. The trick is so obvious it's not even a trick: make two thin doors and laminate them together.

It can't be that easy, you're thinking. And yes, you're right. In many ways you're making two doors, so it's twice the work. The technique works best with woods that have similar rates of movement—apply high-movement hard maple to one side and stable mahogany on the other and the door could develop substantial warping. I used butternut and black locust because their movement rates are similar enough that there shouldn't be much trouble.

Exterior wood doors do take a beating from sun and rain. A Southern-facing door that has no overhang can need refinishing in less than a year, whereas a Northern-facing door with a deep overhang could last for ten years without the need to refinish. This design isn't as stable as solid wood or as stave construction. However, if you use this design to laminate two thin doors together from the same wood, you'll have a door that is more stable than a solid wood one.

I used bolection molding to hold the glass and panels. They stand proud of the surface, giving the door some visual depth. The bolection moldings are applied after you glue up the frame, which is easier to do than integrated moldings. You might look at the muntins over the glass and feel faint—but these are the easiest muntins to make. They are joined with simple laps and floating tenons. The single insulated glass panel is a compromise between a traditional multi-pane look and modern R-value expectations and construction simplicity. While this door takes more time and effort than other doors, the individual steps are easier than the frame-and-panel interior door in **Chapter 7**.

Variations
The proportions, square-profile bolection moldings, rustic-looking locust and screws fit a partly modern, partly Arts-and-Crafts style. Simply changing the any of these will give you a very different looking door. One simple change is to chamfer the muntins for a subtle softening of the lines.

FRAME-AND-PANEL TWO-FACE DOOR

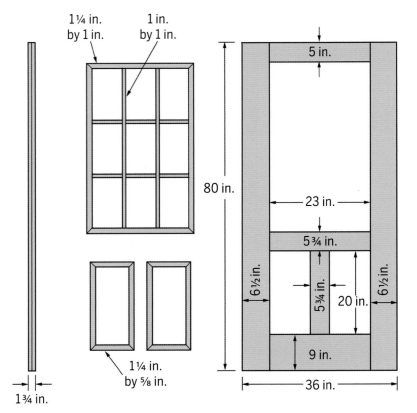

1¼ in. by 1 in.

1 in. by 1 in.

1¼ in. by ⅝ in.

1¾ in.

80 in.

5 in.

23 in.

5¾ in.

6½ in.

5¾ in.

20 in.

6½ in.

9 in.

36 in.

Chamfering a section of the the inside edges of the muntins softens their lines.

Step 1

Mill your rails and stiles for both sides of the door following the steps in **Chapter 2** for **Rough Milling and Fine-Tune Milling** to end up with four stiles, flat and square, that are 6½ in. wide by ⅞ in. thick (or a little under) and at least 80 in. long. At the same time mill the rails to the same thickness, all a little long if possible. They do not need to be perfectly straight. In fact, ⅞ in. thick stiles will be a little flexible and may tend to bow. Mill them flat and smooth on both sides, they will become straight and rigid when you laminate the doors together. If you can't get ⅞ in. thickness out of 4/4 stock, that's all right. The door works just fine with ¹³⁄₁₆ in. or even ¾ in. thick rails and stiles; but the thicker the better.

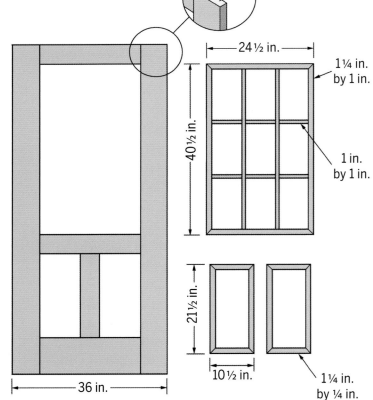

¹³⁄₁₆ in.

24½ in.

40½ in.

1¼ in. by 1 in.

1 in. by 1 in.

21½ in.

10½ in.

1¼ in. by ¼ in.

36 in.

FINISHED DIMENSIONS:
80 in. by 36 in. by 1¾ in. thick

MATERIALS

- Approximately 25 board feet of 4/4 black locust
- Approximately 25 board feet of 4/4 butternut
- 1 insulated tempered glass pane, 38¾ in. by 22¾ in. by ½ in. thick

CUTLIST

- Stiles 4@ 80 in. by 6½ in. by ⅞ in. (2 of each species)
- Top rails 2@ 30 in. by 5 in. by ⅞ in. (1 of each species)
- Lock rails 2 @ 30 in. by 5¾ in. by ⅞ in. (1 of each species)
- Bottom rails 2 @ 30 in. by 10¼ in. by ⅞ in. (1 of each species)

- Mullions 2 @ 25 in. by 5¾ in. by ⅞ in. (1 of each species)
- Panels 4 @ 20 in. by 8½ in. by ⅝ in. (2 of each species)
- Bolection moldings for glass 24 linear feet @ 1¼ in. by 1 in. stock for (half in each species)
- Stock for exterior panel bolection moldings 12 linear feet @ 1¼ in. by ⅝ in. in black locust
- Stock for interior panel bolection moldings 12 linear feet @ 1¼ in. by ¼ in. in butternut
- Stock for muntins 24 linear feet @ 1 in. by 1 in. (half in each species)

HARDWARE

- 3 butt hinges with ball bearings
- Insulated glass pane
- Lockset with handles

Step 2: Lay out the door parts labeling each joint. Do this for both sides.

Step 3: Clamp the four stiles together to lay out the mortises.

Step 2

Lay out both doors to decide the orientation of the parts. Label each joint with letters to keep track of which part goes where.

Step 3

Clamp all four stiles together and lay out the mortises on all at the same time. The mortises for the lock (or middle) rail and the mullions can be the same width as the rails themselves as they do not need shoulders. Only the top and bottom rails need a 1 in. shoulder on the outside edges, and the bottom rail mortise should be split. Clamp the lock rail and the bottom rail together and mark out centered mortises for the mullion or middle stile.

Step 4: Scribe the ½ in. mortises offset from center, closer to the inside face.

Step 5: Measure between your layout lines on the stiles to get the length of your mullions.

Step 5: Dado the tenon faces on the tablesaw, using the rip fence as a width stop. Remember that the tenons are not centered. You will need to change the depth of cut between front tenon faces and back faces.

Step 4: Clamp the thin rails (and stiles) together to make a wide platform for mortising.

Step 4

Cutting the mortises for these doors is a little tricky. Ideally, they would be 5 in. deep and ⅜ in. wide, but I do not have a router bit that thin and long. So I cut them 2¾ in. deep by ½ in. wide because that's the router bit that I have and it still makes a good joint. I do offset the mortise closer to the inside face of the stiles, approximately a heavy ³⁄₁₆ in. and light ⅛ in. from the inside face. When the doors are laminated together the weak inside mortise walls will be supported by each other. Clamp the four stiles together to give the router base support as you cut the mortise on the outside board. Unclamp, reshuffle and reclamp each stile so the one you cut is always on the outside of the stack. This way you don't have to reset your router fence. Cut the mortises for the muntin in the same way. Cut slowly and carefully.

Step 5

Determine the total length of the rails from field measurements, subtracting twice the width of the stiles and adding twice the depth of the mortises. For the muntins, measure between your layout lines and add the mortise depths. Cut your rails to length on the tablesaw using the crosscut fence. Then use a dado blade to cut the tenon faces, guiding the rails with the crosscut fence and the rip fence. Remember that the mortises are not centered in the stiles, so you will need to cut the front faces of all the tenons first, then reset the dado blade depth to cut the back faces of the tenons.

Step 6

Round the tenon ends on the mullions and fit them. Refer to **Steps 14, 16-17** in **Chapter 7** for a more detailed description of fitting the tenons. These are easier as there's no molding to miter.

Step 7

Clamp the mullions between their lock and lower rails and use the assembly to lay out the tenon locations on the lower rail. This is more accurate than relying on measurements as it takes small variations into account. Now trim the shoulders on the upper and bottom rails, then round the tenon ends to match the mortises on all the rails. Fit the tenons into their respective mortises. Take your time and get them perfectly flush, as the fit will affect the glue joint between the two door sides.

Step 8

Dry fit each door, clamping all the joints tight to see how they fit. (Often joints that clamp up fine on their own somehow change when part of the whole door assembly). Disassemble and fix each joint as necessary. Also check the overall squareness of the door by measuring the diagonals. If they are within 1/16 in. you're doing really well. If they are off by more than 1/4 in., you should find the cause and fix it. Remember that you will cut the door into a perfect rectangle after it's assembled. When everything looks good, glue up both doors separately using a waterproof adhesive. I used polyurethane for a little extra time before tack, as there's a mullion to clamp up first.

Step 9

After the glue has cured on both door sides, trim off any glue squeeze out at the joints on both faces. You will need to clean up the

Step 7: With the mullion clamped between, mark the location of the tenon shoulders against the actual mortises.

Step 8: Check the diagonals of the dry fit door sides. They should be within 1/4 in. of each other.

Step 9: Smooth the inside face of both door sides using a block plane and chisel.

Step 10: Laminate the two door sides together, clamping evenly across all surfaces and ensuring the door is flat and not twisted.

Step 11: Trim the excess dried glue from the inside edges of the frame.

inside face of each door side before they are laminated together. Sanding isn't a good option as sanders tend to cut a bit more aggressively at the edges of boards, which will open the glue joint right where you'll see it.

Step 10

Laminate the two door sides together, using waterproof glue. Spread the glue evenly on both faces and clamp them together. I screwed them together through the extra lengths of the stiles at either end to keep them from sliding out of alignment while I clamped them (if you don't have any extra stile length, place your screws along the inner edge of the frames, where the holes will be hidden by the bolection moldings). Clamp up the door on the most flat area in your shop: if your bench top isn't big enough, use the floor or some sawhorses. Shim the clamps as necessary to ensure the door is flat, straight and not twisted, checking by eye and with a level. Use a glue with a slow set to give you time to check and adjust. I used polyurethane, giving me about a half hour to get things right; epoxies can have even longer open times.

Step 11

After the glue has set, remove the clamps and trim away the glue squeeze out on the inside edges. It doesn't need to look pretty or even be perfectly flush, but clear so the glass and wood panels can fit.

Step 12

Cut the door to size (80 in. long by 36 in. wide) with a track saw (or a circular saw and a home made jig). Sand the black locust exterior door face to 220 or 320 grit. Don't worry about the butternut side until later.

Step 13

Mill up black locust stock for the exterior bottom panel bolection moldings, 1¼ in. wide by ⅝ in. thick. I mill the different moldings separately to make it easier to keep track of them and avoid mistakes. (See drawing on facing page)

Step 14

Using the tablesaw cut rabbets on the backside of the locust moldings. Use a regular blade and make two passes, making sure the offcut is on the outside (otherwise you'll get a kickback).

Step 14: Rabbet the bolection moldings on the tablesaw, with the offcut on the outside of the blade.

Step 15: Test the fit of each side of the molding within the door frame.

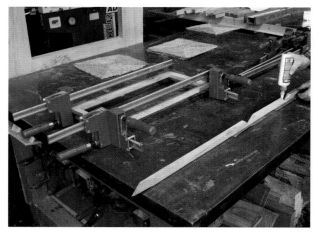

Step 15: Glue up the mitered bolection frame using tape to align the parts and clamps to press the joints together.

Step 16: Glue the bolection frame to the door using a flexible, gap-filling glue such as epoxy.

Molding Profiles

Window bolection molding, interior and exterior.

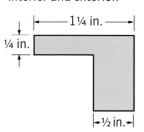

Muntins, interior and exterior.

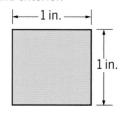

Exterior panel bolection molding.

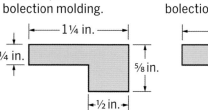

Interior panel bolection molding.

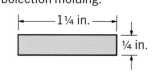

Step 15

Miter the frame members to fit within the panel frames. Glue the frames up separately as if they were a picture frame. Remember to use water resistant glue.

Step 16

When the glue has cured, sand the three sides that show on the bolection molding frame to 220 or 320 grit. Then glue the frames to the door. I used a water resistant epoxy with great flexibility for this, as I expect a little cross-grain movement between the door

Step 19: Find the precise length of the muntins by measuring between the inside corners of the bolection frame members.

Step 20: Align the muntins, evenly spaced, and mark the intersections precisely. Mark the top and bottom on each.

Step 20: Test the fit of the lap joints on scrap pieces to make sure the saw depth and stop blocks are set up accurately.

and the panel frame. When the glue cures, apply a coat of exterior finish or epoxy to the inside of the frame and back of the bolection molding. This is to reduce the chance of rot if rain gets into the panel frame.

Step 17

Peg the joints. See **Chapter 7 Step 26** for detailed description of this process.

Step 18

Mill the black locust bolection moldings for the glass frame, 1¼ in. wide by 1 in. thick. These are thicker than the lower panels to accommodate the thinner glass panel, while still offering a similar appearance. Mold them in the same way and fit them to the frame (but don't glue them up into a picture frame yet).

Step 19

Mill the black locust muntins, 1 in. wide and 1 in. thick. The muntins are captured within the frame, so cut them to length precisely the same as the distance between the interior miters of the frame.

Step 20

Lay out the lap joints on the muntins, spacing them evenly. Mark each joint with a letter on the outward facing sides to help orient you while cutting them. Sand the inside faces of the muntins *before* you cut the joinery (if you cut the joinery first, then sanding will make the joints loose). Cut the laps on the tablesaw, using a stop block to ensure consistency. The depth should be just half the thickness of the muntin, and the width just so you can press the joint together with your fingers. Make test cuts on scrap pieces to set up the blade depth and stop block locations before making the actual cuts. Tradition says the vertical muntins should run top to bottom uninterrupted while the horizontal ones have their lap cut so they fit behind them. If you do it the other way, only traditional woodworkers will notice, though they probably will be rude and say something. We're just that way. Glue up the muntins.

Step 21

Cut floating tenon joints in the ends of the muntins and on the inside faces. I use exterior-grade dominoes made of sipo, 5mm by 30mm, to join the muntins with the frame. Shop made floating tenons will work fine too, just use an exterior grade wood. Glue up the frame and muntins using water-resistant glue. This can take more clamps than it should, considering the mitered corners need clamping as well as the floating tenon joints, so use a slow-setting glue. Sand and finish the frame. This is much easier to do before installation than after.

Step 22

Mill and sand four ⅝ in. thick panels, two in butternut and two in black locust, ¹⁄₁₆ in. shorter than the openings are tall and about ¼ in. less than the openings are wide to allow for expansion and contraction. Finish both of them on both sides.

Step 23

Mill the interior panel bolection moldings in butternut. They should be rectangular in profile–as long as the two panels are flush with the inside face of the door. In this case they do

Step 21: Cut floating tenons on the ends of the muntins and in corresponding places on the bolection moldings.

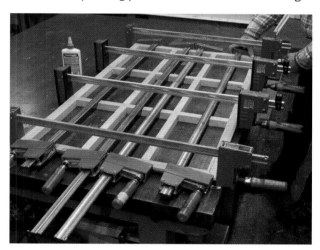

Step 21: Glue up the bolection moldings and the muntins at the same time, ensuring each joint has clamping pressure directly over it, and the corners are clamped from either side.

FITTING THE PANELS

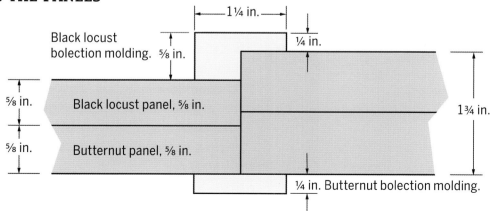

Black locust bolection molding. ⅝ in.

1¼ in.

¼ in.

⅝ in.   Black locust panel, ⅝ in.

1¾ in.

⅝ in.   Butternut panel, ⅝ in.

¼ in. Butternut bolection molding.

Step 24: Drill and countersink pilot holes in the interior panel bolection molding frame.

Step 24: Screw the bolection molding to the frame, pinching the panels in place.

not need a shoulder that fits into the frame (my panels came just flush with the door surface).

If your panels are thinner, add a shoulder to the butternut bolection profile like the black locust molding to pinch the panels tightly in place.

Step 24

Miter the butternut bolection moldings and assemble them into frames, just as for the window bolection moldings. Even though this is interior, I still use a water-resistant glue.

Step 24

Sand the bolection frames after the glue has cured, then drill and countersink screw holes around the outside perimeter of the frame Finish them on both sides then capture the two panels sandwiched in the frame. You can also glue the bolection in place. Unlike glass panels, wood ones rarely break and don't need to be replaced. I use the screws here so they match the ones in the upper frame.

Step 25

Measure and order an insulated and tempered glass pane for the window. It should be ½ in. thick and about ¼ in. smaller in both dimensions than the opening that it fits in (it's very hard to shave glass down to fit, and cutting the frame is not much easier), or about 38¾ in. by 22¾ in., but local measurements should trump these theoretical ones.

Step 26

Set the black locust window bolection frame into the door, gluing it in place with water-resistant glue. After the glue cures, set the insulated glass pane in place with clear silicone caulk. Finally, set the butternut interior frame into the door and screw it in place, just like the panel bolection molding (no glue—if the glass ever breaks, you'll want it to be easily removed).

Step 27

Apply the same high performance exterior finish to the front and back of the door faces, but leave the other edges unfinished until after hanging. Then finish the other edges, except the bottom edge, where you should apply a coat of epoxy, especially to the end grain of the stiles.

Congratulations—you're now the proud parent of a door slab.

Step 27: Apply epoxy to seal the bottom end grain on the door against water.

Steaming Out Dents

Butternut is a lovely wood but dents if you walk past it quickly. Happily, it's easy to steam out dents with an iron. Simply dab the dent with water then press it with a hot iron (cotton heat setting). The iron will boil the water shooting steam into the wood, entering the crushed fibers and popping them back in shape. Do this before you apply a finish, of course, as you'll need to lightly sand afterwards. By the way, sanding out dents doesn't work nearly as well, as you may create raised bumps when you apply a finish.

Steam out dents in the soft butternut with an iron and a little water.

Interior French Doors

12

Windows let in light. Doors let in people. French doors let in light and people. You could say French doors are really large windows on hinges.

Traditional French doors have many small glass panes held in place by muntins. This was because of the limits of glass manufacturing techniques when they were first designed and the look has stuck. Nowadays, modern commercial exterior grade French doors often use a single pane of insulated glass. It's easier to build, offers better R-value and requires less maintenance. You can see how to do this in **Chapter 11,** applying the muntins over the glass. And for an exterior application, this is the superior solution.

For interior applications, the traditional method with "real" glass panes still works better. Single-thickness panes are lighter than insulated glass, allowing you to make a thinner yet durable frame. The trouble is that the muntin framework is relatively difficult and certainly time consuming to make.

As coping and mitering the muntins together in a framework is the only substantial difference from the other frame-and-panel doors in this book, I will focus on that aspect of this project. These muntins are a fussy kind of work—highly satisfying if you enjoy

sharp chisels and precision work involving math. And highly frustrating if you don't.

The step-by-step will describe the making of one door. If you want two, just double the recipe with a few added steps.

INTERIOR FRENCH DOORS

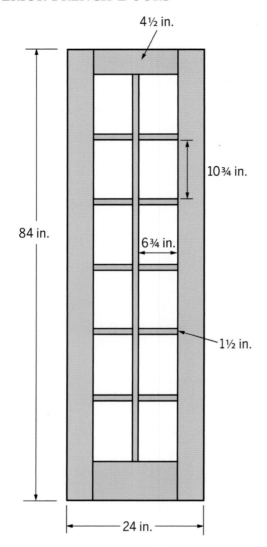

4½ in.

84 in.

10¾ in.

6¾ in.

1½ in.

24 in.

Muntin and Stop Profiles

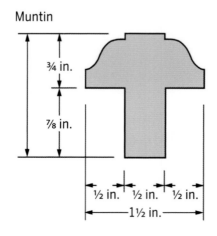

Muntin

¾ in.

⅞ in.

½ in.   ½ in.   ½ in.

1½ in.

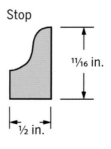

Stop

¹¹⁄₁₆ in.

½ in.

FINISHED DIMENSIONS:
84 in. by 24 in. by 1⅝ in.

MATERIALS
- Approximately 13 board feet of 6/4 cherry
- 10 tempered glass panes, 11⅝ in. by 7⅝ in by ⅛ in.

CUTLIST
- Stiles 2 @ 84 in. by 4½ in. by 1⅝ in. cherry
- Top rail 1 @ 20 in. by 4½ in. by 1⅝ in. cherry
- Bottom rail 1 @ 20 in. by 7½ in. by 1⅝ in. cherry
- Horizontal muntins 5 @ 16½ in. by 1½ in. by 1⅝ in. cherry
- Vertical muntin 1 @ 73½ in. by 1½ in. by 1⅝ in. cherry
- Tempered glass panes 24 @ 11⅝ in. by 7⅞ in. by ⅛ in. thick
- Glass stop, approximately 50 ft. of ½ in. by ⅝ in. cherry (from wider boards)
- Pegs 6 @ 1½ in. by ¼ in. by ¼ in. white oak

HARDWARE
- 2 flush bolts
- Door handle and latch
- 6 @ 3½ in. butt hinges

Milling and Shaping

Step 1

Mill the two stiles and two rails and cut the mortises following **Steps 1-8** in **Chapter 7.** Of course, the dimensions of the parts are different, and you don't have a lock rail, but the steps should be much the same.

Step 2

Mill rectangular stock, 1⅝ in. by 1½ in., enough for the vertical and horizontal muntins. Choose the straightest grain stock you can find for the long vertical muntins and mill it in steps so that it has as little tension left or desire to warp as possible. When you mold and rabbet the muntin stock, you will remove a lot of material which can lead to warping. Mill some extra muntin stock for making practice joints.

Step 3

Mark the locations of the muntins on the rails and stiles, the 1½ in. outside width, on the front edge, the top and the back. This is because you'll cut away most of the lines when you mold and rabbet the edges, and it's good to have the lines already there (they're near impossible to locate accurately after you've cut the moldings, though transferring them around the outside edges of the rails and stiles works well). You can also mark a centered ½ in. wide section at this point, but I generally forget.

Step 4

Mold the profile on the outside face of the door, the entire edge of both stiles and rails (on the inside, the molding is on the window stops). The bit is a ¼ in. radius Roman ogee (it should cut a profile that's ½ in. wide by ¾ in. high). The router table fence should be aligned with the bearing to avoid a bad cut over the mortises. You could cut the molding on the rails and stiles with a hand held router, but it's best to set up a router table for all the moldings as the muntins are too thin to mold safely with a hand-held router.

Step 2: Saw oversize blanks for the muntins on the tablesaw to release the stress. After they rest overnight, joint and plane them to size.

Step 3: Layout the intersections of the muntins on the faces and edges of the stiles and rails.

Step 4: Rout the molding on the edges of the rails and stiles on a router table with a fence.

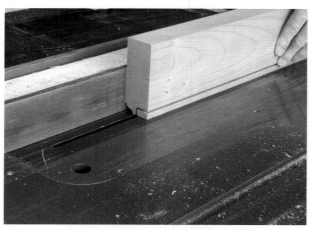

Step 5: Cut the rabbet on stiles and rails using the tablesaw. Keep the waste to the outside of the blade.

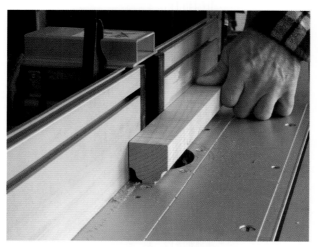

Step 8: Mold the muntin stock on the router table, being extraordinarily careful to apply even pressure downwards to avoid twisting the workpiece away from the fence.

Step 9: Cut the rabbets on the muntins on the tablesaw. Again, be extraordinarily careful to apply even pressure downwards on the workpiece to avoid twisting it away from the fence.

Step 5

Cut a rabbet ½ in. deep and ⅞ in. wide on the inside edge of the rails and stiles using a table saw with either a dado blade or two cuts with a combination blade. It's important to be exact removing the ½ in. as this determines the inside shoulder of your joints. It should be square with the molding on the front face.

Cut the tenons on the rails on a tablesaw with a dado blade. Their length should be just shy of the depths of the mortises from the rabbet (not the molding). **Step 11–12** in **Chapter 7** can give you added direction.

Step 6

Miter the moldings and fit the joints referring to **Steps 14–19** in **Chapter 7** for specifics. Dry fit the frame and ensure all the joints are good and at 90 degrees.

Step 7

Cut the vertical muntin to length, or 1½ in. longer than the distance between the rabbets on the inside (the tenons fit into ¾ in. deep mortises in either rail). Transfer the layout lines for the five horizontal muntins to it from one of the stiles and scribe them around all four sides. Do this now as it will be hard to do accurately after you've molded and rabbeted the muntins.

Step 8

Mold the profile on the outside edges of all the muntins. The muntin stock is small so take care as you mold it. If you apply uneven hand pressure, the workpiece can twist or roll against the fence and ruin the profile.

Step 9

Cut the rabbets on both sides of the muntins as in **Step 5**, leaving ½ in. between them. Again, apply even downward pressure so as to be sure you don't twist the workpiece as you cut the rabbet. Be sure that the waste strip is on the outside of the blade, and not pinched between the blade and the fence.

Step 10: Saw away the waste for the joints on the rails and stiles, well inside your layout lines.

Step 10: Pare the surface of the joint flat and even.

Step 11: Lay out the ½ in. by 1⅛ in. mortises between the rough-cut miters on the rails and stiles.

Step 11: Chop the mortises square and ¾ in. deep with a chisel after drilling out the waste.

Mortising the Muntins into the Rails and Stiles

Before you fit the muntins, cut the mortises in the rails and stiles. You'll use their precise locations to fit the muntins together.

Step 10

Saw away the molding waste on the stiles and rails where the muntins will fit, just as you did for the rails. There's no mortise here yet, but don't get anxious as that comes later. Use a ⅜ in. chisel to pare away the waste.

Step 11

Layout the mortises for the muntins on both the rails and the stiles, ½ in. wide and ¼ in. from the edges. Use the rough edge of the miters as a basic guide, and keep the mortise walls as straight as you can (it's impossible to make a layout line on the roughcut miters, and they shouldn't be finish cut until fitting). Drill out the waste with a ⁷⁄₁₆ in. diameter drill ¾ in. deep, and square up the mortise with a chisel. Put a piece of tape on the drill and the chisel to cut to the right depth. Clamp the stiles together to check that the mortises are evenly spaced.

Step 12

Cut the horizontal muntins to length, adding 1½ in. for the two ¾ in. depth mortises at either end.

Step 13: Lay out the intersections of the muntin miters, ½ in. in from each side of the full muntin width.

Step 14: Use a stop block to center the dadoes in the horizontal muntins.

Step 14: Test the fit of the dadoes on the back of the spine.

Mitering the Muntins

Here is where things get picky. You want to make the lap joints between the muntins as accurately as possible. You might cut a perfectly fitting joint, but it won't do you much good if the distance between them is out of relation to the mortises on the rails and stiles. Number your joints and mark them so you remember which way is up at all times. The work is similar to the lap joints in **Step 15, Chapter 11,** but made more complex with the miters. It helps to think of the process as making a simple lap joint between two ½ in. thick rectangular profile muntins, with a mitered molding as a second step.

You can also cope these joints. It takes more time by hand, but is a little more forgiving. If you're off by a little bit with the miter, you'll have a gap (it can be filled with sawdust and glue). With the cope, most mistakes don't show up as much.

Step 13

If you haven't already, mark the ½ in. wide cuts on the faces of the horizontal muntins and the vertical one, centered between your 1½ in. layout lines.

Step 14

Using a stop block for consistency, dado the horizontal muntins on the tablesaw. A single stop block setting works for all four cuts. Dado the vertical muntin joints by eye. Remember, it's best to cut in small increments. The joint will show on both sides of the door—nowhere to hide. As you can't test each joint's fit before you miter, test the dado width on parts of the spline.

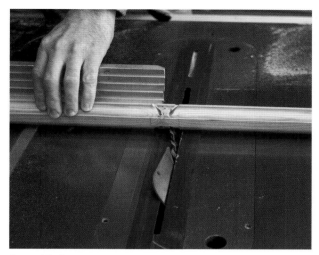

Step 15: Cut the miters on the muntins using the tablesaw and miter gauge with the blade tilted to 45 degrees.

Step 16: Ease the joint together, checking for square and looking from directly above to see where you need to remove material to make it fit.

Step 15

Cut the muntin miters on the tablesaw. You can miter with the miter guide, as in **Step 15** in **Chapter 7**, but I find it particularly difficult to clamp one on the odd-shaped muntin profile. And as the pieces are small, they lend themselves to the tablesaw. This can be difficult with the long one on the tablesaw, so do your best with a handsaw and miter gauge, as you did for the miters in the rails and stiles in **Step 9**.

Step 16

Fit the muntin-to-muntin miters. This is easier said that done, and frankly there's no good word to add on how to fit them well. It's a matter of figuring out what wood to remove that will bring the joint together AND keep everything in alignment. I recommend lots of measuring and rechecking. Start putting the joint together as far as it will go without pushing and you'll see where it touches and where it doesn't. The corners of the top rib in the horizontal muntins are the most fragile: never push when they're touching anything, as you'll dent them and they won't look good in the finished joint. Also, check the squareness of each face of the joint as you fine tune it.

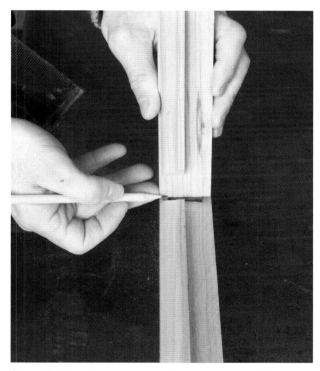

Step 17: Align the uncut muntin tenon against the mortise to mark the shoulder.

Step 17

Miter the ends of the muntins to fit in the rails and stiles in the same way as the muntin-to-muntin joints. These miters are set ¾ in. in from the ends, and tenons cut on the ends. Take your layout lines from the actual tenons

Step 17: Cut the tenon shoulders with a handsaw.

Step 19: Rip the molded stop from a wider board.

you've cut. Cut the tenons with a handsaw and clean them up with a wide chisel.

Step 18

Dry fit the entire door, adjusting the joinery as necessary. Be thoughtful when you clamp, as a too-long muntin will bow if the tenons are too big or too long.

Step 19

Mill the glass stop in lengths of at least 3 ft. and on the edge of a wider board. Then rip the molded edge off on the tablesaw. This is much safer than molding a little bitty stick. In fact,

trying this would be extraordinarily dangerous. The stop should be $^{11}/_{16}$ in. by ½ in. finished. The longer lengths than finished make it easier to sand and finish before mitering to length.

Step 20

Steam out any dents and sand all moldings before assembly up to 220 or 320 grit. With the moldings, I find it easier to clamp them in a vise and use sandpaper wrapped around a dowel, being careful not to round any of the edges or ends. Remember that on the inside, the glass stop will cover most, but not all of the inside edge of the moldings. Festool used to sell a sander that worked on moldings, but they don't anymore, so this is a hand-sanding job. I don't bother sanding the face of the rails and stiles until after assembly.

Step 21

Glue up the muntin-to-muntin joints first with whatever glue you prefer. Then glue up the entire door. You don't have to wait until the muntin-to-muntin joints are dry: just don't twist them if they have started to set. Refer to **Steps 25-28** in **Chapter 7** for details on assembly, pegging and trimming the door to size. Finish the door and the glass stop as you like, though this door has an oil finish as it's interior and does not get much use.

Step 22

Measure the insides of the finished frames and obtain 12 panes of tempered glass the same size. It may be that one or two of the lites are slightly smaller than the rest (it's a mystery how this happens…), so size the glass you buy to the smallest lite. This reduces complexity.

Step 23

Fit the panes with a little clear silicone. This is not for weatherproofing so much as to reduce rattling when you open and close the door.

Step 24

Miter and fit the stop in place over the panes. Miter the ends of the stop on the tablesaw, cutting them so they fit neatly and tightly. Secure them in place with two brads, each about 2 in. from the ends. I don't recommend gluing them in place in case you ever break a pane as it's a *pane* to get them out. Ha ha.

Step 25

If you're making a pair of French doors that meet in the middle, the main added work is the hardware. The main door needs a normal latch. The secondary door houses the strike (treat it just like a jamb, though a thin one). You must be able to secure the strike side of the secondary door to the floor and ceiling, either with flush bolts or with a cremone bolt. To hide the gap between the doors, simply add a nose to the secondary door.

Congratulations—you're now the proud parent of a door slab.

TWO FRENCH DOORS

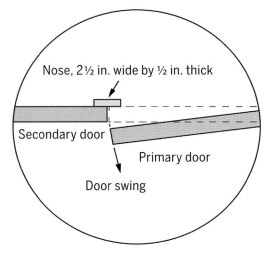

Nose, 2½ in. wide by ½ in. thick

Secondary door

Primary door

Door swing

Flush bolts secure secondary door to floor and ceiling.

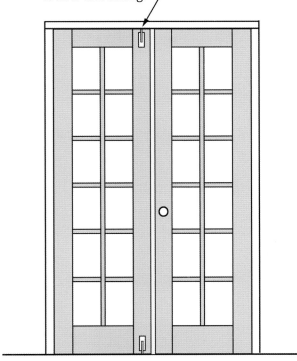

Marbles set between two panes of glass lets colored light through and keeps privacy.

Arts-and-Crafts Glass Panel Door

13

Yes, it's another frame-and-panel door (are you finding there's a pattern? It's because the design works). This one is not the same, however: it is a cross between a French door and a traditional four panel door, gently curved rails and stiles giving the door some character.

As the principles and specifics of making this door are covered in **Chapter 7** and **Chapter 11**, I'll focus here on how to make the curves and curved joinery. No wait! Come back! It's really not that hard, taken step-by step.

Almost all woodworking machines are designed to create and refine straight edges and flat surfaces (e.g. tablesaws, jointers, planers). Many woodworking machines can also handle curves (bandsaws, routers, sanders) but virtually none are specifically designed to cut curves and nothing but them (though jigsaws and coping saws are better at cutting curves that straight lines). This bias leaves most woodworkers "curve blind": uncertain about making them if they wanted to, as it requires adapting tools to the task or using them in new ways.

Templates and routers are perhaps the most common curve-cutting tools, but—surprise—I don't use them to make this door. I do create templates, but refrain from using the router with them. In part this is because I enjoy

the free hand work more and also because working entirely with templates (you need to make two complementary curved templates to reflect both sides of the joint) is more time consuming except if you're making multiples.

Variations

At heart, this project is about learning to use curves. Once you've become comfortable with the basic techniques, a whole new vocabulary of design opens up.

In the rustic knotty oak, it has an Arts-and Crafts feel. Make this same door in a fine-grained wood such as walnut or mahogany, and it begins to have an Art Nouveau feel.

The sharper the radius of the curves, the more difficult it gets to cut the joinery and the more likely you will create short grain or grain runout (in which the run of the grain does not follow the edge but intersects it). Short grain is far more liable to crack and break, so use it in decorative, not structural elements.

The glass panes can be made in wood. The curved design will become less noticeable, though, unless you use a contrasting wood for the panels. I used to mix woods this way but found it very rarely produced a pleasing design, so try it with caution (the more the woods are alike, the better

Arts-and-Crafts Glass Panel Door

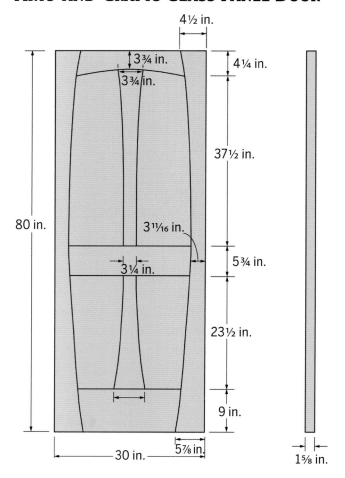

4½ in.

3¾ in.

3¾ in.

4¼ in.

80 in.

37½ in.

3¹¹⁄₁₆ in.

3¼ in.

5¾ in.

23½ in.

9 in.

30 in.

5⅞ in.

1⅝ in.

Glass Pane Profiles

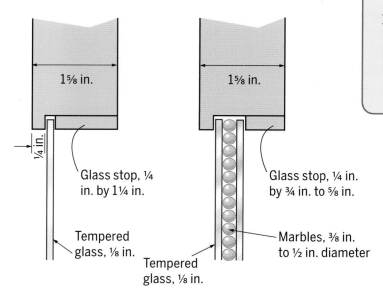

1⅝ in.

¼ in.

Glass stop, ¼ in. by 1¼ in.

Tempered glass, ⅛ in.

1⅝ in.

Glass stop, ¼ in. by ¾ in. to ⅝ in.

Tempered glass, ⅛ in.

Marbles, ⅜ in. to ½ in. diameter

Finished Dimensions:
80 in. by 30 in. by 1⅝ in.

Materials
- Approximately 16 board feet of 8/4 red oak for the rails and stiles.
- Tempered glass panes (or lights)

Cutlist
- Stiles 2 @ 80 in. by 5⅞ in. by 1⅝ in. red oak
- Top rail 1 @ 24½ in. by 4¼ in. by 1⅝ in. red oak
- Lock rail 1 @ 24½ in. by 5¾ in. by 1⅝ in. red oak
- Bottom rail 1 @ 21¼ in. by 9 in. by 1⅝ in. red oak
- Upper mullion 1 @ 41 in. by 3¾ in. by 1⅝ in. red oak
- Lower mullion 1 @ 27½ in. by 5 in. by 1⅝ in. red oak
- Four ⅛ in. thick tempered glass panes to template shape
- Approximately 30 linear feet of glass stop, ¼ in. by 1 in. red oak
- Pegs 8 @ 1½ in. by ¼ in. by ¼ in. white oak

Hardware
- 3 @ 4 in. butt hinges
- Bronze lizard handle
- Bullet catch

(i.e. a maple frame with birds eye maple panels or white oak with red oak panels)

With clear glass, the door does not offer any privacy. Seeded glass will partially obscure visibility and frosted glass entirely. Another option is to capture glass marbles between two panes of glass, for a rainbow color effect when light shines through it.

Step 1

Using a scrap of ¼ in. plywood, or other sheet good, make a template blank for the curve that's on the stiles (you'll also use it for the mullions and the top rail). Rip the piece 5¾ in. wide 80 in. to 82 in. long, same dimensions as your stiles. If you use a longer piece, use a square to mark the ends of the stile at 80 in.

This is a gentle curve, an arc with an approximately 44 ft. radius. You could use a 50 ft. tape measure or length of wire and set it somewhere down a hallway or outside, but there are easier ways. Some woodworkers like the precision of math, plotting points along the curve: and it works, but I can't be bothered. I simply set three clamps, one at each end and one in the middle to bend a long straight edge between them. My longest straight edge is 6 ft., so I lay out the curve in two parts. If you don't have one, you can also use a thin strip of wood—anything long that bends easily and doesn't have any kinks in it. Use the measurements in the drawing to locate the clamps. To make sure your curve is even, sight it from one end, looking down its length.

An option is to make your curve with a ¾ in. thick scrap of MDF or cabinet grade plywood. This template will allow you to cut the curve on the stiles using the template as a router guide. I don't do it this way as the template has to be perfect and I make too many mistakes with the router that are hard to fix (a slight wobble will create a void).

Step 1: Bend a long straight edge between clamps to scribe the gentle curved edge of the stiles template.

Step 2

Make curves for the upper and lower mullions, using the measurements in the drawing. There are two ways to do this: make two separate templates, bending the curves as you like. I tend to flare the curve a little at top and bottom and keep them straighter in the middle, a bit like a lily. The other option is to use the stile template, aligning the smaller mullion parts along the corresponding parts of the stile template.

Step 3: Cut away the waste just shy of the template line on the bandsaw.

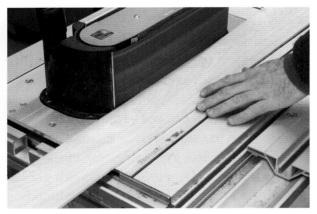

Step 4: Sand to the template line on a stationary belt sander.

Step 6: Lay out the curves on the stiles with a pencil, clamping the template at either end.

Step 3

Cut the templates on a bandsaw, staying just outside the line. You can also cut them with a jigsaw or handsaw.

Step 4

Sand to the line on the templates using a stationary belt sander. Take your time here and work the sander to get an even curve. Sight down the curve to adjust any uneven spots. If you don't have a stationary belt sander, try a hand file: the work will go quickly with ¼ in. thick plywood.

Step 5

Mill your rails, stiles and mullions and lay out your door parts following **Steps 1–2** in **Chapter 7** (and **Chapter 2** for **Rough Milling** and **Fine-Tune Milling**. All the parts should remain rectangular at this time, and through the steps cutting the mortises—the curves come later. You can lay out the locations of the rails on the edges and faces of the stiles, but don't bother to lay out the mortises quite yet.

Step 6

Lay out the curve on your stiles, and those on the mullions while you're at it, clamping the template at either end and aligning the back of the template with the outer edge of the stile.

Step 7

Rip the top ends of the stiles to 4½ in. width, stopping the cut before you get near the layout line for the curve. Cut the strip off with a handsaw. This allows you to cut deeper mortises in the lock and top rails.

Step 8

Mark the length of the mortises in the stiles and on the rails for the mullions. They should not be the full width of the rails, but ⅜ in. in from either side to allow for a ¼ in. rabbet for the glass panels. At the ends of the stiles, they should be at least an inch from the top and bottom. The large bottom rail should

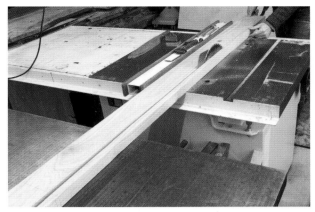

Step 7: Rip the upper section of the rail to 4½ in. wide stopping the cut before the stile curve widens.

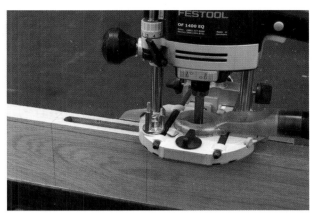

Step 9: Mortise the stiles for the rails 2½ in. deep or deeper with a router and fence.

Step 10: Saw the curve on the stiles on a bandsaw, staying just shy of your layout line.

Step 11: Sand the curve edge of the stiles on a stationary belt sander to the layout lines, as evenly as possible.

be divided into two tenons, ½ in. between them. Mark the width of the mortises with the marking gauge. Scribe from both sides to ensure you've centered the mortise. The mortise should be a little less than half the thickness of the stile, about ¾ in. or ¹³⁄₁₆ in.

Step 9
Cut the mortises using a router and fence, following the instructions in **Steps 5-7, Chapter 7**.

Step 10
Cut the curve on the stiles and mullions using a bandsaw, or jigsaw if you prefer, staying just shy of your layout line. Try to make sure that the cut is as square as possible to the face of the stile.

Step 11
Sand the curved edge of the stiles and mullions on a stationary belt sander. Here, it is key to sand as flat and square an edge as possible.

Step 12
Determine the actual total length of the rails, i.e. the length that shows plus a tenon at either end. This might seem a bit trickier with these stiles, as they're curved, but it isn't. Do remember, though, that the lock rail and top rail will be one measurement while the bottom rail will be another, so you have to repeat this process twice: measure the depth of the mortise and subtract it from the width of the stiles **at the same point** (this gives you the distance from the bottom of the mortise to the outside edge of the stile). Multiply that

Step 13: Scribe the curved stile shoulder on the face of the rail with the parts aligned at 90 degrees.

Step 15: Cut the faces of the tenons on the tablesaw with a dado blade, angling the miter gauge to follow the line of the joint. Cut up to the curved layout line as close as you can, but not over it.

Step 14: Transfer the end of the tenon shoulder to the other side, after marking out the tenon on the face of the rail.

Step 16: Chop the waste up to the curved tenon shoulder layout line with a chisel.

by two, then subtract it from 30 in., the total width of the door. This equals the total length of the rail. Crosscut the rails to length.

Step 13

Lay out the tenon shoulders on the rails by placing the stiles on top and scribing the slightly curved stile edge onto the rail. First align the rail at 90 degrees to the outside edge of the stile, then align the end of the rail with the bottom of the mortises.

Step 14

Flip the tenon and do the same for the other side. Mark the shoulders of the tenons

on the edges of the rails to make it easier to align the two faces of the tenon.

Step 15

Mark and cut the tenons on the table with a dado blade. Set a dado blade to the height of the tenons, and cut just shy of your line. You want the tenons to be just a bit too thick for the mortises so you can fit them precisely by hand. The mullions are all fit against rails at 90 degrees (except the top rail); but you'll need to angle the miter gauge by about 5 degrees for the bottom rail, and a few degrees less for the top rail. Cut close to, but not over, these curved layout lines on the face of the rail. The dado will cut a straight edge, leaving material at either end of the curved joint, and that's fine for now. Use the rip fence as a stop to get even results on either side of the 90 degree joints. However, the rip fence is less useful for the angled dados.

Step 16

Trim the tenon shoulders to your curved layout line with a chisel and mallet. Again, your work does not have to be perfect at this point, but approximate the curve as best you can without cutting over the layout line. The shoulder should be cut square or slightly undercut.

Fitting Joinery

As in **Step 7** of **Chapter 11**, fit the mullions to the rails first. Here, you have upper and lower mullions, so you'll need 8 ft. clamps to fit them at once. Use this dry-fit assembly to lay out the shoulders of your tenons for the stiles at one time, paying close attention to how straight and square they fit. You can also lay out the tenons for each joint separately, then negotiate the discrepancies when you dry fit the whole door.

Step 17

Lay out the shoulders of the mullion tenons by aligning them with their rails. Cut the waste from the tenons on the bandsaw. You can also do this work with a backsaw and coping saw. Round the tenon corners to fit the mortises and trim the tenon shoulders square. Fit the tenons.

Step 18

With the entire mullion-rail assembly clamped up, straight and true (sometimes this is harder than it seems—often easily solved by releasing pressure on the clamps), align the stiles along side.

Step 19

Lay out the shoulders of the tenons for the stiles. Make sure each rail intersects the

Step 16: Check the squareness of the tenon shoulder, and cut straight or undercut as necessary.

Step 19: Mark the tenon ends with a square, registered off the back edge of the stile, and the rail aligned at 90 degrees with the stile.

Step 20: Trim the tenon ends off on the bandsaw.

stile at 90 degrees using a framing square. Use a small square on the outside edge of the stile to help guide your pencil in the right location. Then take apart the tenon-rail assembly to fit each joint individually.

Step 20

For these tenons, now cut the waste from the tenons on the bandsaw. Round the tenon corners to fit the mortises and trim the tenon shoulders square.

Step 20: Pare the tenon shoulders even or slightly undercut with a chisel.

Step 22: Mark where the tenon shoulder touches the stiles with the joint dry fit. These are the spots you will cut away.

Step 21: Test the fit of the tenons against the mortise shoulders.

Step 21

Fit the tenons to the stiles the same way as for any square tenon, checking the thickness of the mortise wall against the amount removed from the tenon. See **Steps 16–17** in **Chapter 7** for more detail.

Step 22

Check the curved connection and pencil where the joint touches (where you need to cut away). Make sure the joint is aligned at 90 degrees before deciding what to cut. Don't go too far

with the shoulder lines or you'll end up with a crooked joint or a smaller door. If you do go over the lines, then do it to the other rails as well to ensure they're all the same relative width between the stiles. To make a smooth curved edge along the joint face, use a shoulder or block plane to even up the chisel cuts. It is helpful if the shoulder is slightly undercut and you have a little to shave with the plane. Repeat this process and check the fit of the joint until you're satisfied with the seam on all the joints.

Step 23

Dry clamp the whole door and check that the joints are still good. If any joints are out (*how could they be! They were perfect when clamped up on their own!*), chances are good you need merely trim one end of the tenon to let the rail slide left or right a bit to make it right. Annoyingly it is true—individual perfection rarely translates to group perfection: the latter requires negotiation.

Step 24

Glue up the door with a slow-set, low-tack glue such as polyurethane. Give yourself time to align everything, tap it into place as necessary. Of course, start with the mullions and rails, lastly add on the stiles. Check diagonals and if you're out, remember that you can square up the door after assembly by trimming the edges.

Step 22: Smooth the tenon shoulders with a sharp shoulder plane.

Step 25

After the glue has set, rough sand both sides to remove any glue squeeze out and any unevenness from the joints.

Step 26

Rout the rabbet for the glass panes with a router and ¼ in. deep rabbeting bit. If your bit won't cut the entire 1 ⅜-in. depth at once, make several passes at greater depths. This is not a deep rabbet, but for a simple glass pane that doesn't move, it's enough.

Step 27

The router will leave curved corners which you should chop square, as in **Step 15** in **Chapter 14**. Saw and/or chop the curved corners square. This isn't the traditional way to do it: it should rather be done like the door in **Chapter 7,** with the rabbet cut before the tenons are made and the square front edge treated like a molding and mitered at the corners. However, this also works and is much easier.

Step 28

Make a cardboard or plywood template for the glass panes. Glass shops are sometimes highly accurate, and sometimes not. To guard against this, make two copies of each template (they often lose the one you give them) and make the templates on the small side, so if the panes they give you are too big, they may still fit the door. Get the glass made.

Step 29

Peg the joints, finish sand the door, and trim the edges parallel and square, and finish the door with oil. For specifics, refer to **Steps 26–29** in **Chapter 7.** 1½ in. pegs are long enough here, though longer is fine just try not to hammer all the way through.

Step 30

Make and finish the glass stop. At ¼ in. thick, it should be flexible enough to press along the curves.

Step 31

Fit the glass panes with a bead of clear silicone along the inside edge of the frame. Let this cure. It's enough to hold the glass just fine, but fit the stop on the inside for a finished look. Do not miter the corners. Instead, fit the vertical pieces first, (clamping them in place until you can secure them with brads), then the horizontal pieces last, as in **Step 24** in **Chapter 14.** This gives the illusion of the right intersection lines. If you're using brads to secure the stop, be careful hammering near the glass. Pneumatic brads or small screws work a little better. Just don't glue it in place or you'll have a real mess if the glass ever breaks.

Congratulations—you're now the proud parent of a door slab.

Screen Door

14

Not every open door lets everything in. Not every closed door keeps everything out. Dutch doors are designed so the top half can open leaving the bottom shut, letting air and bugs in, but keeping animals out. Screen doors are designed to let air in, but no bugs, and when they're shut. Got it?

Screen doors are traditionally thin and lightweight. In part this is because they don't need to be thicker. Screening weighs next to nothing so the door is only holding its own frame up. They also need to be thin to fit the outside of a jamb already occupied by a full-size door. The trouble with thin, though, is obvious—thin does mean flimsy and making a durable screen door is something of a trick.

Of course (you know it's coming), I use pegged mortise-and-tenons and stock at least ⅞ in. but preferably 1 in. thick, to make a durable screen door that looks nice and holds up over time. You can make thinner doors, but you'll need to come up with a different way to integrate the screen, perhaps losing the stop that covers the spline.

I made this screen door out of western red cedar, a lightweight yet rot and weather resistant softwood. It is sold in my area as dimensional building lumber in housing construction-oriented lumberyards, and not in hardwood lumberyards.

A turnbuckle across the frame can help support frail joinery.

Variations

There are simpler ways and faster ways to make screen doors, which is to say with exterior grade floating tenons or dominoes. If you go this route, I highly recommend adding a turnbuckle to keep the door joinery together. As screen doors effectively cover the exterior doors, they are often as decorative if not more ornate than the main door. Let your whimsy be your guide to curved corner brackets, carvings and other details.

Screen Door

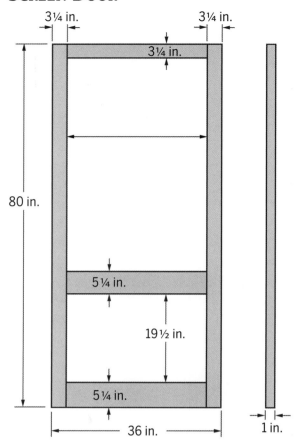

3¼ in. 3¼ in.

3¼ in.

80 in.

5¼ in.

19½ in.

5¼ in.

36 in.

1 in.

Fitting a Screen Door

Screen doors need to be thin and have offset hardware to fit in the same jamb as the main door.

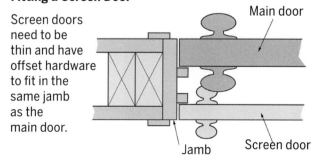

Main door

Jamb

Screen door

Kerf for Screening

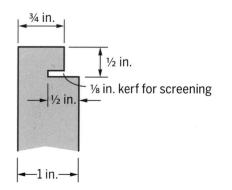

¾ in.

½ in.

½ in.

⅛ in. kerf for screening

1 in.

Finished Dimensions:
80 in. by 36 in. by 1 in.

Materials

- 1 @ 2 x 4 by 14 ft. clear western red cedar
- 1 @ 2 x 6 by 9 ft. clear western red cedar
- 1 @ 1 x 6 by 6 ft. clear western red cedar
- 18 square feet of screening
- 18 linear feet of spline

Cutlist

- Stiles Stiles 2 @ 80 in. by 3¼ in. by 1 in. clear western red cedar
- Rails 2 @ 34½ in. by 5¼ in. by 1 in. clear western red cedar
- Rails 1 @ 34½ in. by 3¼ in. by 1 in. clear western red cedar
- Panel @ 29⅜ in. by 10½ in. by ⅜ in. clear western red cedar
- Stop 2 @ 60 in. by ½ in. by ¼ in. clear western red cedar
- Stop 2 @ 30½ in. by ½ in. by ¼ in. clear western red cedar
- Pegs 6 @ ⅞ in. by ¼ in. by ¼ in. white oak

Hardware

- Screen door latch
- 2 @ 2½ in. butt hinges

Step 1

Mill your rails and stiles following the steps in **Chapter 2** for **Rough Milling and Fine-Tune Milling** to end up with two stiles, straight, flat, square that are 3¼ in. wide by 1 in. thick and at least 80 in. long. As you're working with 2 x 4 material, you can make a door up to about 1⅜ in. thick if you prefer and have room for it in the jamb, but 1 in. is thick enough. At the same time mill the rails to the same thickness, all a little long if possible. As they are shorter, they may not need to be fine-tune milled the same number of times, but be sure they end up the same thickness as the stiles and just as straight and square.

Step 2: Orient the rails and stiles and mark the corresponding joints

Step 3: Mark the locations of the rails on the stile edges at the same time.

Step 5: Scribe mortises 7/16 in. wide centered on the stiles.

Step 6: Cut the mortises 2⅜ in. deep on a slot mortiser.

Step 2

Lay out the door parts in the orientations that look best and mark the joints for reference. Label each joint with letters to keep track of which part goes where.

Step 3

Clamp the stiles together and lay out the mortises on both at the same time. Mark where the rails intersect the stiles.

Step 4

Mark the length of the mortises in the stiles. Around the screen area, they should be about ¼ in. in from either side and around the lower panel, about ⅝ in. At the ends of the stiles, they should be at least ¾ in. from the top and bottom of the stiles.

Step 5

Mark the width of the mortises with the marking gauge. Scribe from both sides to ensure you've centered the mortise. Make the mortise between ⅜. in. and ½ in. wide.

Step 6

Cut the mortise 2⅜ in. deep, or as deep as your tooling can go. For this door I used a slot mortiser because the stock is thin and light. If you don't have a slot mortiser, a router will do a perfectly good job. Just clamp the stiles together and even an additional board, to give the router base enough support in the cut. See **Step 5** in **Chapter 7** for details on cutting mortises with a router.

Step 7: Rout a ½ in. bead on the panel-facing edges of the rails, balancing the router carefully.

Step 8: Scribe the tenon shoulders with a sharp marking gauge to get a clean thin line.

Step 9: Cut the tenon faces on the tablesaw with a dado blade.

Step 7

Rout the outward-facing edges of the rails with a ½ in. bead (except the very top and bottom, of course). Don't rout the inside edges as you'll just cut it away to add the stop over the screening groove. You can do this with a hand held router, just take care to keep it steady on the rail edges for a smooth cut. Making this cut on a router table is another good option. This is a nice detail and adds no complexity to the construction. If you mold the stiles as well, it's a different story (see **Chapter 7** on how to miter an integrated molding).

Step 8

Measure and mark the tenon shoulders on the rails. If you've milled the stiles perfectly 3½ in. wide, and cut the mortises 2⅜ in. deep, then the total length of the rails should be 34¼ in. But of course, you should calculate this on your own.

Step 9

Cut the tenon faces on the tablesaw with a dado blade. I first mill the tenon faces too thick on one of them, then take small increments off and test the fit each time. When the tenon fits snugly with only hand pressure, I've got a good fit. This is an alternate and faster way to fit tenons. It can be accurate depending on how accurate your mortises are. It doesn't accommodate variation, though, the way the technique shown in **Steps 16-17** in **Chapter 7** does. I take this shortcut here because an uneven joint face can be sanded out easily in the soft cedar.

Step 10

Trim the tenon shoulders and round the tenon sides to fit each mortise. If necessary adjust the tenons to align with the pencil marks on the stiles. Leave the pencil marks there for now, but erase them before assembly.

Step 10: Pare the shoulders with a very sharp chisel as softwood tends to compress under dull blades.

Step 11: Check the diagonals in the dry fit frame.

Step 12: Rout a ⅜ in. wide groove for the panel spaced closer to the back of the door.

Step 13: Rout a ½ in. wide rabbet for the screen around the inside face of the door.

Step 11

Dry fit the frame and check the diagonals and joint connections to see that everything is aligned at 90 degrees and fits well.

Step 12

With the frame still clamped up, rout the groove for the panel, ⅜ in. wide but not centered. Locate the groove approximately ¼ in. from the back of the door as the bead weakens the front edge. Square up the groove corners with a chisel to make space for the rectangular panel. Measure the frame for the panel size, subtracting ¹⁄₁₆ in. from the length and ¼ in. from the distances to allow for wood movement.

Step 13

Lastly before you take the frame apart, rout the shallow ¼ in. deep by ½ in. wide rabbet for the screen stop.

Step 14

Cut the groove for the spline on either a router table or the tablesaw, stopping the cut at either end on the stiles (on the rails, you can groove the tenons without weakening them much). As the groove is ⅛ in. wide a standard tablesaw blade is the right thickness, but it is hard to make a stopped cut on the tablesaw. Also the blade diameter means you'll have to finish a fair amount of the groove with a chisel. Consequently, it is easier to do on a router table with a grooving bit. Cut the groove ½ in. deep from the face of the

Step 16: For tight-fitting joinery, use a waterproof PVA glue to assemble the frame.

Step 17: Saw the corners of the rabbet for the screen square with a fine-toothed saw.

Step 18: Hammer in ¼ in. by ¼ in. oak pegs from the back side of the door, being very careful to avoid setting them too deep in the soft cedar.

door. Eventually, the ends of the groove and shallow rabbet will need to be squared up, but I wait until after assembly to do that.

Step 15

Cut and sand the panel and test the fit in the groove.

Step 16

Assemble the door frame with a water-resistant glue (unless this is an interior screen door, or the screen is really protected from the elements—no, no "unless." Just use waterproof glue). It is not a good idea to glue the panel in place. However, if you use an exterior grade plywood panel, you can glue it in the frame and add to the door's strength.

Step 17

After the glue has cured, square up the corners of the screen groove and the screen stop rabbet. Use a ⅛ in. chisel for the groove and a fine-toothed saw and wider chisel to cut the rabbet. Make sure your chisels are sharp and use a shearing cut movement or they will crush rather than cut the soft cedar fibers.

Step 18

Peg each joint from either the inside or the outside, depending on where you'd like to see the pegs. The pegs can be made of any durable wood, though I have a preference for white oak, ⅞ in. long and ¼ in. square. See **Step 26** in **Chapter 7** for more detail on making and setting pegs. As the cedar is soft, it is easy to pound the peg too far such that it either comes out the other side or compresses the cedar fibers such that you can see the peg location from the other side. There's no good way to fix this except drilling out the peg and setting a longer one that goes through both sides.

Step 19

Sand both door faces, steaming out any dents, of which you'll have many as the cedar is soft (for details on steaming out dents, see the sidebar on page 107 in **Chapter 11**).

Step 20

Trim the top and bottom of the door square and to 80 in. height.

Step 21

I finished the door with exterior grade oil that leaves a matte finish. I think its a good pairing with the soft cedar, which will dent and age quickly, giving the door an older rustic look in no time. Apply several even coats to the cedar as it can blotch if you apply the oil unevenly.

Screening

Fitting a screen is both easy and maddening. You can do it with standard shop tools, but the process is made easier with a "screen mouse" or spline wheel. This tool has a thin wheel with a concave face that puts even pressure on the spline as you push it into the groove.

Step 22

Spread the screen material over the frame on your bench. The screen material should be wide and long enough to stretch at least two inches over the groove you've cut for the spline. The more overlap the better as it will prove handy later. Although not necessary, I like to tape it at the corners to keep the screen aligned. You may also want to clamp the door to the table to keep it from moving around while you work.

Step 23

Start pressing the spline into the groove at a corner, capturing one edge of the screen. You can use any tool without sharp edges that fits the groove. I use a screwdriver tip with rounded edges, but a spline wheel works better and even a pizza cutter with a dull blade can

Step 23: Press the spline over the screening into the groove. Exert even pressure to avoid kinking or bunching.

Step 24: Cut the screen at a 45 degree angle into the corners to keep the screen from overlapping there.

work fine (be careful not to cut the screen with a sharp pizza cutter). It's important to press the spline in evenly to avoid kinking the screen, so run the screwdriver along the spline in a fluid motion. Check to see that the screen stays aligned with the door, i.e. that the mesh stays square to the frame.

Step 24

When you reach a corner, cut the screen at a 45-degree angle so they don't overlap under the spline.

Step 25: Tighten the screen as needed by tugging from the outside and rolling the spline back into the groove with a dull screwdriver tip.

Step 26: Trim the excess screening with a knife, keeping it to the outside edge of the spline.

Step 27: Nail ½ in. by ¼ in. screen stop over the spline with evenly spaces brass nails to give the interior a finished look.

Step 25

After you've set the spline around all four sides, the screen may be loose in places. To tighten it, pull on the outside edge and roll the spline back down into the groove with the screwdriver.

Step 26

Trim the excess screening with a knife, careful not to cut either the spline or the screen on the inside edge of the spline.

Step 27

Mill ¼ in. thick by ½ in. wide strips from the leftover cedar for screen stop. Cut the vertical strips to length first, then nail them in the stile using evenly-spaced copper weatherstripping nails. Next, set the horizontal strips in the two rails. These aren't necessary to hold the screening in place, but they give the interior a finished look.

Congratulations—you're now the proud parent of a screen door slab.

Step 28

A note on the location of knobs or handles. It's a very common mistake to set them at the same height as the front door handles, then find the handles hit each other when you hang the door. Set the screen door handles a few inches lower or higher than the main door to avoid this problem. I tend to set them lower, but I'm not sure why other than habit.

Gallery

City Palace, Udaipur, Rajasthan, India, 19th century: open frame construction with vertical boards secured on back side.

City Palace, Udaipur, Rajasthan, India, 19th century: Bronze door mimicking wood frame construction.

City Palace, Udaipur, Rajasthan, India, 19th century: Asymmetric wood frame-and-panel door with metal overlay.

New Hampshire, USA, 19th century: frame-and-panel with handle 24 in. from floor.

Kanazawa, Japan, modern: Cedar frame-and-slat-panel exterior sliding doors.

Kanazawa, Japan, modern: Cedar frame door with exterior-grade plywood panel.

Kanazawa, Japan, modern: Cedar mitered-frame with fretwork internal panels and circular bolection molding.

Kanazawa, Japan, modern: Frame construction door with natural bamboo slat paneling.

Connecticut, USA, modern: Maple frame-and-panel door with natural edge casing.

Connecticut, USA, early 20th century: Pine board-and-batten door.

New York, USA, modern: Walnut frame-and-panel doors.

Connecticut, USA, modern: Pine internal-frame with vertical boards and centered handle.

Connecticut, USA, modern: Frame-and panel construction with plywood panels and applied moldings

Connecticut, USA, modern: Butternut internal frame with vertical boards.

Connecticut, USA, modern: Oak internal-frame with vertical boards and strap hinges.

(right) **Connecticut, USA, modern:** Cherry frame-and-panel bedroom doors.

Connecticut, USA, modern: Cherry exterior French doors with cremone bolt.

Connecticut, USA, 19th century: Pine board-and-batten exterior door (inside face).

Connecticut, USA, 19th century: Pine board-and-batten exterior door outside face).

Nantucket, USA, 19th century: frame-and-ornate panel and fretwork construction.

Nantucket, USA, 19th century: frame-and-seven-panel construction.

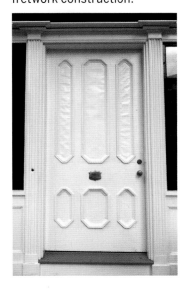

Nantucket, USA, 19th century: frame-and-two-panel construction with bolection moldings.

(far left)
Nantucket, USA, 19th century: frame-and-tin-panel construction and bolection moldings.

(left)
Nantucket, USA, 19th century: frame-and-panel with four lites construction.

Connecticut, USA, modern: mahogany frame-and panel construction with blown glass lites.

London, England: Traditional frame construction but with center handle hardware, separate lockset on the edge, and a molding along the bottom edge to deflect water.

Added moldings can dress up a plain frame and panel door.

Oxford: Medieval style linenfold carved door that is rectangular, set into non-rectangular frame, and same with curved top.

About the Author

My first woodworking job was in 1986 at Wood Interiors by Rodger Reid in New Preston, Connecticut, where I worked summers while attending college. Rodger specialized in solid wood interior paneling and cabinetry. He and Ed Downs taught me most of the trade. My first job was to move a giant pile of lumber from one place to another, then back again after I'd rebuilt the rack. My last job was building five solid bird's eye maple frame-and-panel doors for one of Bill Cosby's New York townhouses. I wrote "chicken heart proof door" on the inside of one.

After a flirtation with academia, I served as an editor at the Taunton Press from 1996 to 2000, first at *Fine Woodworking* magazine, then in the book department. Since then I've written tool reviews, articles and *Traditional Box Projects*. During this time, I replaced all the doors in my house, including the front and back doors. They are all still working, in spite of raising two kids through teenage-hood.

In January 2000, I started my own business designing and making furniture, cabinetry and doors. My first shop was in our two-car garage. It was roomy but had 7 ft. ceilings. Naturally, two of my earliest commissions were bookcases, both over 9 ft. tall. I had to build them horizontally and assemble them outside in good weather. This spurred me to build a shop with 11 ft. ceilings in 2002. Naturally since then, customers have wanted mostly tables, beds, front doors, buffets, the occasional kitchen, even music stands and salad bowls—but nothing taller than 7 ft, including doors. This is one of those universal laws to which I humbly submit.

Index